An introduction to linear and nonlinear scattering theory

π Pitman Monographs and Surveys in Pure and Applied Mathematics 78

An introduction to linear and nonlinear scattering theory

G F Roach

University of Strathclyde

CRC Press is an imprint of the
Taylor & Francis Group, an **informa** business
A CHAPMAN & HALL BOOK

First published 1995 by Longman Group Limited

Published 2019 by CRC Press
Taylor & Francis Group
6000 Broken Sound Parkway NW, Suite 300
Boca Raton, FL 33487-2742

First issued in paperback 2019

ISBN-13: 978-0-367-44826-4 (pbk)
ISBN-13: 978-0-582-09230-3 (hbk)

Visit the Taylor & Francis Web site at
http://www.taylorandfrancis.com

and the CRC Press Web site at
http://www.crcpress.com

AMS Subject Classifications: (Main) 35P25, 47A40, 34B25
(Subsidiary) 35L05

British Library Cataloguing in Publication Data

A catalogue record for this book is
available from the British Library

Library of Congress Cataloging-in-Publication Data

A catalog record for this book is available

ISSN 0269-3666

Contents

Preface

This monograph has two main purposes. First, to act as a companion volume to more advanced texts by gathering together the principal mathematical topics commonly used in developing scattering theories and, in so doing, provide a reasonably self-contained introduction to linear and non-linear scattering theory for those who might wish to begin working in the area. Second, to indicate how these various aspects might be applied to problems in mathematical physics and the applied sciences. Of particular interest will be the influence of boundary conditions.

The intention is to present the material so that it is just as persuasive to mathematicians interested in the spectral analysis of initial boundary value problems for partial differential equations as to applied scientists who, understandably, might require rather more quantitative results from their mathematical model. However, this does present certain difficulties since the mathematical literature on scattering theory is based very much on functional analysis and the theory of operators in Hilbert spaces, two topics which are not always to be found in the curricula for non–mathematicians, whilst the applied science literature in the area is somewhat more inclined towards heuristic arguments and formal manipulations. This book is meant to provide a bridge between these two extremes. Throughout the emphasis is on concepts and results rather than on the fine details of proof. The proofs of results simply stated in the text, many of which are lengthy and very much of a technical nature, can be found in the references cited in the Commentary. Briefly, this book can be thought of as providing a basic tool kit for the spectral and asymptotic analyses of wave propagation phenomena.

Each chapter will begin with a summary indicating the underlying physical concepts and the results to be expected from the theory to be presented. In addition, worked examples will be introduced, whenever this can reasonably be done, to illustrate theory and its applications. These two aspects coupled with the self–contained nature of the development will, it is hoped, make the material accessible and indeed attractive to both the applied scientist and the mathematician and in so doing persuade them to read more widely in this research area.

Naturally, in an account such as this there can be no real novelty in the subject matter itself. The results described in this book represent the work of many authors over a number of years. Since scattering theory is such an active area of research it would be almost impossible to compile a complete bibliography in the field. Neverthless, mention must be made of the pioneering works of Lax and Phillips [41], Pearson [56], Reed and Simon [60], Segal [70], Strauss [75] and Wilcox [85]; the influence of the works of these authors has been considerable and is gratefully acknowledged. In particular, a profound debt of gratitude is owed to Rolf Leis and Calvin Wilcox who have been such an inspiration over the years.

I would also like to express my gratitude to the many colleagues with whom I have had such useful discussions. In particular I would thank Wilson Lamb, Pauline McCullagh, Desmond McGhee and John Paul Rogan who have read the manuscript, in its various forms, and offered so many helpful suggestions.

A special word of thanks must also go to Mary Sergeant for the patient way in which she tackled the seemingly endless task of typing and proof corrections.

Finally, I want to express my appreciation to Longmans Publishing Staff for their efficient and friendly handling of the publication of this book.

Glasgow GFR

1994

Chapter 1
Introduction and outline of contents

Time dependent equations occur frequently in mathematical physics with perhaps the most common examples being the linear wave equation and the linear heat equation. When studying the initial value problems and initial boundary value problems associated with such equations the first requirement is to declare what type of solution to such problems is being sought. For example, are the solutions expected to be classical solutions in the sense that they posses the appropriate number of continuous derivatives or are they expected to be more generalised types of solution? Once existence and uniqueness results have been obtained for the required type of solution attempts can then be made to obtain other details of the solutions such as their regularity and their asymptotic behaviour.

In this book we shall be particularly interested in the time-asymptotic behaviour of solutions to time-dependent problems posed in unbounded regions which often involve a boundary. That is, we shall be concerned with the development of so-called scattering theories.

Scattering theory means different things to different people. Broadly speaking it can be thought of as the study of the interaction of an evolutionary process with a non-homogeneous and possibly non-linear medium. Certainly, it has played a central rôle in mathematical physics over the years with perhaps the earliest investigation of such phenomena being attributed to Leonardo da Vinci who studied the scattering of light into the geometrical shadow of an opaque body. Subsequently, other scattering processes have been discovered and investigated in such varied fields as acoustics, quantum mechanics, medical diagnosis and non–destructive testing processes.

The evolution of a system can be modelled in terms of an initial boundary value problem the solution of which describes the state of the system at any point and at some future (or past) time. Specifically, given initial data, f, the state u, of the system at any point x and time t can be written, abstractly in the form

$$u(x,t) = U(t)f(x)$$

where $U(t)$ is an operator describing the time evolution of the system.

In principle therefore, using the above expression, we can obtain the state of the system at all points (x, t). However, in practice $U(t)$ is difficult to compute. Indeed, such a computation, which in many practical situations can soon become virtually unmanageable because of its complexity, is not always something which can be justified bearing in mind the needs of the experimentalists.

This situation can be eased by making recourse to the results and techniques of scattering theory which have been so successfully developed and exploited for problems in quantum mechanics.

Experiments in quantum mechanics are characterised by the fact that the system undergoes a measurement at a time t_o and also at a later time t_1 and that for most of the time interval $(t_1 - t_o)$ there is negligible interaction between the various components of the system. Indeed components only interact for a small part of the interval $(t_1 - t_o)$. This means that for the greater part of the experiment the component parts are behaving as though they were "free". For this reason, at least, it is unnecessary to solve completely the initial boundary value problem which models the system in order to relate theoretical predictions to the available experimental results.

This philosophy and the associated techniques can be applied to problems in many fields other than quantum mechanics. In this connection we would particularly mention the pioneering works of Lax and Phillips (1967) and Wilcox (1975).

Scattering theory, as developed in quantum mechanics, is associated, almost entirely, with perturbations of the Laplacian interpreted as an operator in $L_2(\mathbf{R}^n)$. This so-called **potential scattering** is now well developed and quite well understood. However, for **target or obstacle scattering** the situation is by no means as satisfactory.

In this book we shall indicate how scattering theories can be developed in fields other than quantum mechanics and survey the results, particularly for target scattering, which have been obtained recently.

First, we clarify what was meant by saying that a component in the system was behaving as though it were 'free". This is perhaps most readily achieved by considering the motion of two particles. The trajectory of the k-th particle, $k = 1, 2$, can be described by a three dimensional vector-valued function $\mathbf{r}_k(t)$ which represents a curve in $\mathbf{R}^3$. In classical mechanics a particle is in free motion if its trajectory is defined by

$$\mathbf{r}(t) = \mathbf{r}_o + \mathbf{v}t, \quad \ddot{\mathbf{r}}(t) = 0.$$

Consequently, when the two particles are a large enough distance apart, so that their interaction can be regarded as virtually negligible, their motion can be considered as being almost along a straight line $\mathbf{r}_k^-(t)$ with uniform velocity $\mathbf{v}_k^-$, $k = 1, 2$.

During the time that particles are close they interact and their motion is, in general, highly complex. Eventually when they move apart, they cease to interact and once again their motion can be considered as being almost that of a free particle in the sense that they move almost along some straight line, $\mathbf{r}_k^+(t)$, with uniform velocity $\mathbf{v}_k^+$, $k = 1, 2$.

This approximation process can be made precise in a natural way by requiring, for $k = 1, 2$

$$\lim_{t \to -\infty} | \mathbf{r}_k(t) - (\mathbf{r}_k^- + \mathbf{v}_k^- t) | = 0$$
$$\lim_{t \to -\infty} | \dot{\mathbf{r}}_k(t) - \mathbf{v}_k^- | = 0$$
$$\lim_{t \to +\infty} | \mathbf{r}_k(t) - (\mathbf{r}_k^+ + \mathbf{v}_k^+ t) | = 0$$
$$\lim_{t \to +\infty} | \dot{\mathbf{r}}_k(t) - \mathbf{v}_k^+ | = 0.$$

The vectors $\mathbf{r}_k^-, \mathbf{v}_k^- (r_k^+, \mathbf{v}_k^+)$ characterise a free state of the k-th particle called the **incoming (outgoing) asymptotic state** of the k-th particle. One of the main aims of scattering theory is to show that these incoming and outgoing asymptotic states actually exist. In this particular case, for example, if long range interaction is being studied then no such states exist. However, when the above limits do exist then we say that the particles are **asymptotically free** as $t \to \pm\infty$.

The problems we shall be concerned with here are those which are modelled mathematically in terms of some differential expression together with certain associated boundary and initial conditions which help in characterising its domain of definition. Scattering processes associated with such problems can then arise as a consequence of perturbing either the symbolic form of the differential expression, so–called **potential** scattering, or the domain of definition of such an expression, so–called **target scattering**, or indeed both. As might be expected from their names, such perturbations occur naturally and frequently in applications.

In developing a scattering theory, a comparison is made between the solutions of a perturbed system, the so–called hard problem, and the solutions of an unperturbed system, the so–called easy or free problem, the solutions of the latter being readily obtained. The asymptotic behaviour of the two systems is investigated as the time t tends to $\pm\infty$ with the aim of determining whether or not the perturbed system appears to behave like the unperturbed system in the distant past and/or in the distant future, that is, whether or not it is an "asymptotically free" system.

To illustrate the different notions appearing in scattering theory frequent reference will be made to the following initial boundary value problem for the forced wave equation in one space dimension.

$$\Box u(x,t) := \left(\frac{\partial^2}{\partial t^2} - c^2\frac{\partial^2}{\partial x^2}\right)u(x,t) = f(x,t) \quad x > 0, |\, t\,| < \infty \tag{1.1}$$

$$u(x,0) = u_0(x), \quad x > 0 \tag{1.2}$$

$$u_t(x,0) = u_1(x), \quad x > 0 \tag{1.3}$$

$$u(0,t) = g(t), \quad |\, t\,| < \infty \tag{1.4}$$

When $f \equiv 0$ the unforced initial value problem (1.1) to (1.3) has a solution which can be represented in the form

$$u(x,t) = \frac{1}{2}\{u_0(x-ct) + u_0(x+ct)\} + \frac{1}{2c}\int_{x-ct}^{x+ct} u_1(s)ds. \tag{1.5}$$

The result (1.5) is the celebrated d'Alembert's solution of the one-dimensional wave equation.

We notice that in order for the solution given in (1.5) to be such that $u \in C^2(\mathbf{R}\times\mathbf{R})$ then the initial data must satisfy

$$u_0 \in C^2(\mathbf{R}), \quad u_1 \in C^1(\mathbf{R}).$$

Furthermore, it can readily be shown that for such initial data the solution given by (1.5) is uniquely determined by the initial data and, moreover, the solution is independent of any boundary conditions which might be imposed.

A solution to the forced initial value problem (1.1) to (1.3) can be obtained by means of Duhamel's Principle and is given by

$$u(x,t) = \frac{1}{2}\{u_0(x-ct)+u_0(x+ct)\} + \frac{1}{2c}\int_{x-ct}^{x+ct} u_1(s)ds + \frac{1}{2c}\int_D f(s,\tau)dsd\tau \tag{1.6}$$

where D is a triangle with vertices at the points (x,t), $(x-ct,0)$, $(x+ct,0)$.

The representations (1.5) and (1.6) clearly indicate the manner in which the given data evolves into the required solutions. Consequently, a comparison of the solutions to the forced (perturbed) system and to the unforced (unperturbed) systems could then be centred on a discussion of the integral

$$\frac{1}{2c}\int_D f(s,\tau)dsd\tau. \tag{1.7}$$

However, a more efficient means of comparison and one which can be applied to a wide range of problems will be indicated later.

An alternative approach which can be adopted when studying the initial value problem (1.1) to (1.4) is to set

$$u_t(x,t) = v(x,t).$$

The initial boundary value problem (1.1) to (1.4)then reduces to the first order system

$$\begin{bmatrix} u \\ v \end{bmatrix}_t - \begin{bmatrix} 0 & 1 \\ -L & 0 \end{bmatrix}\begin{bmatrix} u \\ v \end{bmatrix} = \begin{bmatrix} 0 \\ f \end{bmatrix}. \tag{1.8}$$

$$\begin{bmatrix} u \\ v \end{bmatrix}(x,0) = \begin{bmatrix} u_0(x) \\ u_1(x) \end{bmatrix}. \tag{1.9}$$

where L represents the differential expression $(-c^2\partial^2/\partial x^2)$. We would simply remark at this stage that the boundary condition (1.4) is accounted for in the domain of definition of the differential expression L. The details of this accommodation will be given later. With this understanding (1.8), (1.9) can be written in the form

$$w_t + Gw = F, \quad w(0) = w_0 \tag{1.10}$$

where

$$w = \begin{bmatrix} u \\ v \end{bmatrix}, \quad G = \begin{bmatrix} 0 & -1 \\ L & 0 \end{bmatrix}, \quad F = \begin{bmatrix} 0 \\ f \end{bmatrix}.$$

This partial differential equation can also be interpreted, by means of the calculus of vector-valued functions, as an ordinary differential equation. This technique will be properly introduced in the next chapter so, with this understanding, we shall write (1.10) as

$$\frac{dw}{dt} + Gw = F, \quad w(0) = w_o \tag{1.11}$$

where w is understood to be a function of the variable x for each fixed value of t. We emphasise that in (1.11) G depends only upon x whereas F might depend on both x and t. Therefore, for the moment we shall think of G simply as a constant and use the familiar integrating factor technique to write (1.11) in the form

$$\frac{d}{dt}[e^{tG}w] = e^{tG}F.$$

Integrating we obtain

$$[e^{\xi G}w(\xi)]_s^t = \int_s^t e^{\xi G}F(\xi)d\xi.$$

Therefore, the initial boundary value problem for equation (1.1) in which the initial data given at $t = s$ rather than at $t = 0$ has a solution given by

$$w(t) = U(t-s)w(s) + \int_s^t U(t-\xi)F(\xi)d\xi \tag{1.12}$$

where

$$U(t) = e^{-tG}.$$

Yet another means of investigating the problem (1.1) to (1.4) is once again to use the calculus of vector-valued functions but now to write (1.1) to (1.4) in the form

$$\frac{d^2u}{dt^2} + Lu = f, \quad u(0) = u_0, \quad u_t(0) = u_1 \tag{1.13}$$

where L is the differential expression introduced above. Solving (1.13) we find that, formally at least, the solution of (1.1) to (1.4) can be written in the form

$$u(x,t) = \cos(tL^{\frac{1}{2}})u_0(x) + (L^{-\frac{1}{2}}\sin L^{\frac{1}{2}})u_1(x) + H(x,t;f) \tag{1.14}$$

where $H(x,t;f)$ denotes a particular integral of (1.13).

Once mathematical respectability has been given to these approaches then we shall find that the problems (1.11) and (1.13) and their associated solution representations

(1.12) and (1.14) respectively can conveniently be taken as the starting point for investigating both linear and non-linear scattering problems.

In developing a scattering theory we study the time-asymptotics of a given evolutionary system by comparison with a simpler, free, evolutionary system, the solutions of which are assumed to be readily obtainable.

To see how this comparison can be made and the asymptotic behaviour determined consider an unperturbed or free system governed by the initial value problem

$$\left(\frac{\partial^2}{\partial t^2} - c^2\frac{\partial^2}{\partial x^2}\right) u_1(x,t) = 0 \tag{1.15}$$

$$u_1(x,0) = u_{10}(x), \qquad u_{1,t}(x,0) = u_{11}(x)$$

and a perturbed system governed by initial value problem

$$\left(\frac{\partial^2}{\partial t^2} - c^2\frac{\partial^2}{\partial x^2} + q(x)\right) u_2(x,t) = 0. \tag{1.16}$$

$$u_2(x,0) = u_{20}(x). \qquad u_{2,t}(x,0) = u_{21}(x).$$

If now we introduce the expressions

$$G_1 := \begin{bmatrix} 0 & -1 \\ L & 0 \end{bmatrix}, \quad G_2 := \begin{bmatrix} 0 & -1 \\ L-q & 0 \end{bmatrix} \tag{1.17}$$

$$U_1(t) := \exp(-tG_1), \quad U_2(t) := \exp(-tG_2). \tag{1.18}$$

and argue, formally, in a manner similar to that indicated above, then we see, on recalling (1.12), that the required solutions are given by

$$w_1(t) = U_1(t)\phi_1, \quad w_1(0) = \phi_1 = \begin{bmatrix} u_{10} \\ u_{11} \end{bmatrix} \tag{1.19}$$

$$w_2(t) = U_2(t)\phi_2, \quad w_2(0) = \phi_2 = \begin{bmatrix} u_{20} \\ u_{21} \end{bmatrix} \tag{1.20}$$

where $w_j,\ j = 1,2$ are as in (1.12) and $\phi_j,\ j = 1,2$ denotes the appropriate initial data.

A means of comparing these two solutions can be developed using the structure of Hilbert spaces. Specifically we consider expressions of the form

$$\| w_2(t) - w_1(t) \|$$

where $\| \cdot \|$ denotes a norm on the chosen Hilbert space. We then find that

$$\begin{aligned} \| w_2(t) - w_1(t) \| &= \| U_2(t)\phi_2 - U_1(t)\phi_1 \| \\ &= \| \phi_2 - U_2^*(t)U_1(t)\phi_1 \| =: \| \phi_2 - \Omega(t)\phi_1 \| . \end{aligned}$$

where * denotes the adjoint operator in the chosen Hilbert space structure.

We shall see in the next chapter that the last step is always possible provided the $U_j, j = 1, 2$ can be interpreted as unitary operators in the chosen Hilbert space.

To remove the time dependence in this comparison and in so doing provide a means of discussing the asymptotic behaviour of the solutions we take the limit as $t \to \pm\infty$ to obtain

$$\lim_{t \to \pm\infty} \| w_2(t) - w_1(t) \| = \| \phi_2 - \Omega_\pm \phi_1 \| \tag{1.21}$$

where

$$\Omega_\pm := \lim_{t \to \pm\infty} \Omega(t) = \lim_{t \to \pm\infty} U_2^*(t)U_1(t). \tag{1.22}$$

If it can be shown that the two solutions exist and that the various steps leading to (1.21) can be justified then it will still remain to prove that the limits in (1.22), the so-called wave operators, actually exist and, ideally, are invertible. When all this has been achieved then we see that if the initial data for the perturbed and free problems are related according to

$$\phi_1 = \Omega_\pm^{-1} \phi_2, \tag{1.23}$$

then the limit in (1.21) is zero thus indicating that the perturbed problem is asymptotically free as $t \to \pm\infty$. That is, solutions of the perturbed problem with initial data ϕ_2 are time asymptotically the same as solutions of the free problem with initial data ϕ_1 given by (1.23).

Consequently, if solutions to the two systems are known to exist in the form (1.19) and (1.20) then, keeping (1.23) in mind, we would expect there to exist elements $\phi_\pm$ such that

$$w_2(t) \sim U_1(t)\phi_\pm \quad \text{as } t \to \pm\infty. \tag{1.24}$$

The $\pm$ is used to indicate the possibly different limits as $t \to \pm\infty$.

We would emphasise that it is not automatic that both the limits implied in (1.24) should exist. That is, the solution w_2 could be asymptotically free as $t \to +\infty$ but not so as $t \to -\infty$.

Combining (1.20) and (1.24) we have $U_2(t)\phi_2 \sim U_1(t)\phi_\pm$ that is

$$\phi_2 \sim U_2^*(t)U_1(t)\phi_\pm = \Omega(t)\phi_\pm.$$

Thus we conclude that

$$\Omega_\pm : \phi_\pm \to \phi_2. \tag{1.25}$$

Although the above discussion has been almost entirely formal nevertheless it has been sufficient to indicate the two fundamental problems of scattering theory. First the question of the **existence** of the **wave operators** $\Omega_\pm$. Secondly there is the matter of so-called **asymptotic completeness.** This amounts to an investigation of whether or not **all** solutions of the perturbed system are indeed asymptoticallly free as $t \to \pm\infty$ and is closely related to the existence of such elements as $\phi_\pm$. There are a number of ways of addressing these two problems and perhaps the two most notable are the Schrödinger operator method and the Lax-Philips method. In the former the problem is reduced to the comparison of certain operators on two Hilbert spaces whilst in the latter certain subspaces of a Hilbert space, the incoming and outgoing subspaces, are studied. Whichever means is adopted we must make the above developments more precise mathematically. To do this two requirements are immediate. First, we should introduce the elements of operator theory into the discussion. This will provide a natural framework within which a problem can be given a sound mathematical formulation. Second, we should provide an interpretation of expressions such as $\exp tG$ and $L^{\frac{1}{2}}$. This will require the use of results from the spectral theory of operators and from the theory of semigroups of operators. Broadly speaking we shall use semigroup theory to obtain existence and uniqueness results and spectral theory to develop constructive methods.

Since this book is meant to be an introductory text it is felt that it should be as self-contained as possible. The intention is to give, at a leisurely pace, a reasonably comprehensive overview of the various processes associated with the development of scattering theories which can act as a preparation for progressing to more specialised and demanding texts. With this in mind a Commentary on the various Chapters can be found at the end of the book. This will provide historical background, references and a guide to further reading. The various topics are arranged as follows.

We introduce in Chapter 2 a number of concepts and results from functional analysis and operator theory which are used regularly in subsequent chapters. The presentation is mainly restricted to giving basic definitions, formulating the more important theorems and illustrating the results with examples. More advanced topics in analysis than those appearing here will be introduced as required. This will have the virtue of emphasising their particular role in the development of a scattering theory.

In Chapter 3 we are concerned with outlining, in a slightly less formal manner than previously, more of the strategies which can be adopted in developing a scattering theory. This will be done by means of two examples, the first centred on target scattering whilst the second deals with certain aspects of potential scattering. The final section outlines how scattering theories for more general problems can be developed.

In finite dimensional spaces a linear operator can be represented in terms of matrix with respect to a basis of associated eigenfunctions. A generalisation of this notion to an infinite dimensional space setting is complicated by the fact that the spectrum of an operator can now consist of more than simply eigenvalues. To ease the situation we introduce in Chapter 4 the Spectral Theorem and indicate how it can provide the required generalisations of the spectral representation of an operator and the spectral decomposition of a space with respect to an operator.

In Chapter 5 we introduce the elements of semigroup theory. We shall indicate how this will enable questions of existence and uniqueness of solution to be settled in a systematic manner.

The wave operators are properly introduced in Chapter 6 where we also outline some of their more important properties. This is done most conveniently in an abstract setting; some particular cases are considered later.

The remaining Chapters are concerned with the application of the various techniques and strategies which have been introduced so far. In Chapter 7 we begin the investigation of how an incident wave field is distorted in the presence of a bounded obstacle. In contrast to potential scattering, which is essentially an initial value problem, attention must now be paid to boundary conditions. Chapter 8 provides an indication of how a linear scattering theory can be developed. In Chapter 9 we are concerned with showing how the techniques involved in developing a linear scattering theory can be adapted to accommodate nonlinear problems. The final Chapter gives a commentary on the work outlined in previous chapters together with suggestions for further reading.

Chapter 2
Analytical preliminaries

2.1 Introduction

The mathematical structure of scattering theory is developed within the framework of so-called linear spaces. These linear spaces involve generalisations of the familiar concepts which were introduced in elementary courses in vector algebra and the analysis of functions of a real variable. In these early courses we invariably worked with the value of the function at some point in its domain of definition rather than with the abstract quantity of the function itself. This strategy will no longer be adequate for our purposes. However, in order to be able to work with such abstract elements we shall need generalisations of concepts such as addition, multiplication by a scalar, magnitude and distance. The required generalisations are obtained by extending the familiar notions of vector algebra and geometry in $\mathbf{R}^n$. We shall find that this leads quite naturally to the notions of a metric, a generalisation of the familiar Euclidean distance function in $\mathbf{R}^n$, a metric space and a normed linear space. Furthermore, it will no longer be sufficient to work in finite dimensional spaces such as $\mathbf{R}^n$ and $\mathbf{C}^n$ but instead we must be prepared to work in infinite dimensional spaces. This we shall be able to do very conveniently in the structure of so-called Hilbert spaces.

In this chapter we indicate these various generalisations. In doing so we prove as little as possible. Our main aim here is to collect together and illustrate by examples the various concepts, basic definitions and key theorems, with which the reader might not be familiar, which we shall use frequently in subsequent chapters.

2.2 Preliminaries

We begin with the concept of a **vector space**. This will provide a framework within which abstract quantities can be manipulated algebraically in a meaningful manner.

Definition 2.1

A **vector space (linear space)** over a set of scalars $\mathbf{K}$ is a non-empty set X of elements $x, y, \ldots$ called **vectors** together with two algebraic operations called vector addition and multiplication by a scalar which satisfy

1. $(x+y)+z = x+(y+z)$, $x, y, z \in X$.

2. There is a zero element $\theta \in X$ such that $x + \theta = x, \quad x \in X$.
3. If $x \in X$ then there exists $(-x) \in X$ such that $x + (-x) = \theta, \quad x \in X$.
4. $x + y = y + x, \quad x, y \in X$.
5. $(\alpha + \beta).x = \alpha.x + \beta.x \quad \alpha, \beta \in \mathbf{K}, x \in X$.
6. $\alpha.(x + y) = \alpha.x + \alpha.y, \quad \alpha \in \mathbf{K}, x, y \in X$.
7. $\alpha.(\beta.x) = (a\beta).x, \quad \alpha, \beta \in \mathbf{K}, x \in X$.
8. There is an unit element $1 \in \mathbf{K}$ such that $1.x = x, \quad x \in X$.

For ease of notation we shall suppress the dot in 5 to 8. Furthermore, we would remark that in the following $\mathbf{K}$ will usually be either $\mathbf{R}$ or $\mathbf{C}$.

We shall often need the notion of distance between abstract quantities. This we can introduce by mimicking the familiar processes in Euclidean geometry.

Definition 2.2

A metric space, M, is a set X and a real valued function d, called the **metric** or **distance function**, defined on $X \times X$, such that for all $x, y, z \in X$,

(i) $d(x, y) \geq 0$.

(ii) $d(x, y) = 0$ if and only if $x = y$.

(iii) $d(x, y) = d(y, x)$.

(iv) $d(x, z) \leq d(x, y) + d(y, z)$.

The last relation is known as the **triangle inequality**.

We would emphasise that a given set X can be made into a metric space in different ways simply by employing different metric functions d. For clarity we shall sometimes denote a metric space as (X, d) in order to make explicit the metric employed.

Example 2.3

Let $X = \mathbf{R}^n$ and d be the usual Euclidean distance function. That is, if $x = (x_1, \cdots, x_n)$

and $y = (y_1, \cdots, y_n)$ are any two points in $\mathbf{R}^n$ then the distance between them is

$$d(x,y) = \left\{ \sum_{i=1}^{n} (x_i - y_i)^2 \right\}^{\frac{1}{2}}$$

It is this distance function and its properties which we mimicked in formulating Definition 2.2. Clearly (X, d) is a metric space.

Example 2.4

Let $X = C[0,1]$, the set of continuous real-valued functions defined on [0,1]. This set can be made into two metric spaces, M_1 and M_2, where

$$\begin{aligned} M_1 &= (X, d_1) \\ d_1(f,g) &= \max_{x \in [0,1]} \mid f(x) - g(x) \mid \\ M_2 &= (X, d_2) \\ d_2(f,g) &= \int_0^1 \mid f(x) - g(x) \mid dx. \end{aligned}$$

In each case we must of course show that d_1, d_2 satisfy (i) to (iv) in Definition 2.2. In these cases (i) to (iii) are obviously satisfied whilst (iv), usually the hardest property to establish, is satisfied by virtue of the well known properties of the modulus and of Riemann integrals.

The notion of distance which we have just introduced allows us to give a meaning to the word convergence which is applicable to abstract quantities.

Definition 2.5

A sequence of elements $\{x_n\}_{n=1}^{\infty}$ in a metric space $M = (X, d)$ is said to converge to an element $x \in X$ if

$$d(x, x_n) \to 0 \quad \text{as } n \to \infty.$$

In this case we often write either $x_n \to x$ as $n \to \infty$ or $\lim_{n \to \infty} x_n = x$ where it is understood that the limit is taken with respect to the distance function d.

Different metrics can induce different convergence results. For instance in Example 2.4 we have

$$d_2(f,g) \leq d_1(f,g), \quad f, g \in X$$

and therefore if we are given that a sequence $\{f_n\}_{n=1}^{\infty} \subset X$ is such that

$$f_n \to f \in X$$

with respect to d_1 then it follows that also

$$f_n \to f \in X$$

with respect to d_2. However, if we are given that the sequence converges with respect to d_2 then it does not follow that the sequence also converges with respect to d_1.

Definition 2.6

A sequence $\{x_n\}_{n=1}^{\infty}$ of elements of a metric space (X, d) is called a **Cauchy sequence** if for all $\epsilon > 0$ there exists $N(\epsilon)$ such that $n, m \geq N(\epsilon)$ implies $d(x_n, x_m) < \epsilon$.

The following result is standard.

Theorem 2.7

In any metric space $M = (X, d)$ every convergent sequence

(i) has a unique limit

(ii) is a Cauchy sequence.

It must be emphasised that the converse of (ii) is false. There is a possibility that a Cauchy sequence in (X, d) might not converge to any limit element $x \in X$. However, this difficulty can be avoided by confining attention to certain preferred metric spaces.

Definition 2.8

A metric space in which all Cauchy sequences converge is called a **complete** metric space.

It can be shown for the metric spaces M_1, M_2 in Example 2.4 that M_1 is complete but M_2 is incomplete.

This indicates that what we really need to do is to enlarge the given set X by adding to it limits of all possible Cauchy sequences in X. The original set X would then be contained in some larger set $\tilde{X}$ which would have the required properties. The connection between X and $\tilde{X}$ is contained in

Definition 2.9

Given a metric space $M = (X, d)$ a set $Z \subset X$ is said to be **dense** in X if every element $y \in X$ is the limit, with respect to d, of a sequence of elements in Z.

To indicate that this enlargement or completion as it is called can in fact be made we need

Definition 2.10

A function f from a metric space (X_1, d_1) to a metric space (X_2, d_2) is **continuous** if, for $\{x_n\}_{n=1}^{\infty} \subset X_1$,

$$f(x_n) \to f(x)$$

with respect to the structure of (X_2, d_2), **whenever** $x_n \to x$ with respect to the structure of (X_1, d_1).

Definition 2.11

Let $M_j = (X_j, d_j)$, $j = 1, 2$ be metric spaces. A function f which satisfies

(i) $f : X_1 \to X_2$, one–to–one and onto (bijection)

(ii) preserves metrics in the sense that

$$d_2(f(x), f(y)) = d_1(x, y), \quad x, y \in X_1$$

is called an **isometry** and M_1, M_2 are said to be **isometric.**

It is clear that an isometry is automatically a continuous function. Furthermore isometric spaces are essentially identical as metric spaces in that any result for a metric space $M = (X, d)$ will also hold for a metric space which is isometric to it.

With this preparation we can now state a result which will indicate in what sense an incomplete metric space can be made complete.

Theorem 2.12

If $M = (X, d)$ is an incomplete metric space then it is possible to find a complete metric space $\tilde{M} = (\tilde{X}, \tilde{d})$ so that M is isometric to a dense subset of $\tilde{M}$.

The familiar concepts of open and closed sets on the real line extend to arbitrary metric spaces according to the following definition

Definition 2.13

If $M = (X, d)$ is a metric space then

(i) the set

$$B(y;r) := \{x \in X : d(x,y) < r\}$$

is called the **open ball** in M, of radius $r > 0$ and centre $y \in X$.

(ii) A set $G \subset X$ is called **open** (with respect to d) if for all $y \in G$ there exists $r > 0$ such that

$$B(y;r) \subset G.$$

(iii) A set $N \subset X$ is called a **neighbourhood** of $y \in N$ if

$$B(y;r) \subset N \quad \text{for some } r > 0.$$

(iv) A point x is called a **limit point** of a subset $Y \subset X$ if

$$B(x,r) \cap \{Y \setminus \{x\}\} \neq \phi \quad \text{for all } r > 0.$$

(v) A set $F \subset X$ is called **closed** if it contains all its limit points.

(vi) The union of F and all its limit points is the **closure** of F, denoted $\bar{F}$.

(vi) A point $x \in Y \subset X$ is an **interior** point of Y if Y is a neighbourhood of x.

A particularly important type of metric space is the so-called normed linear space.

Definition 2.14

A **normed linear space** is a vector space X over $\mathbf{R}$ or $\mathbf{C}$ together with a mapping $\| \cdot \|: X \to \mathbf{R}$, known as a norm on X satisfying

(i) $\| x \| \geq 0$ for all $x \in X$.

(ii) $\| x \| = 0$ if and only if $x = \theta$, the zero element in X.

(iii) $\| \lambda x \| = | \lambda | \, \| x \| \;\; \forall \, \lambda \in \mathbf{K}, x \in X$.

(iv) $\| x + y \| \leq \| x \| + \| y \|$ (triangle inequality).

The pair $(X, \|\cdot\|)$ is referred to as a real or complex normed linear (vector) space depending on whether the underlying field is **R** or **C**.

Example 2.15

$X = \mathbf{R}^n$ is a real normed linear space with a norm defined by

$$\| x \| = \| (x_1, \cdots, x_n) \| = \left\{ \sum_{k=1}^{n} | x_k |^2 \right\}^{\frac{1}{2}} .$$

Example 2.16

$X = C[0,1]$ is a real normed linear space with a norm defined by either

$$\| f \|_\infty = \sup_{x \in [0,1]} | f(x) |$$

or

$$\| f \|_1 = \int_0^1 | f(x) | \, dx.$$

We notice that any normed linear space $(X, \|\cdot\|)$ is also a metric space where the metric function d is defined by

$$d(x, y) = \| x - y \| .$$

This is the so-called induced metric on X. With this understanding we see that such notions as convergence, continuity, completeness, open sets and closed sets which were introduced above for metric spaces carry over to normed linear spaces. Typically we have

Definition 2.17

Let $(X, \|\cdot\|)$ be a normed linear space. A sequence $\{x_n\}_{n=1}^{\infty} \subset X$ is said to converge to $x \in X$ if given $\epsilon > 0$ there exists $N(\epsilon)$ such that

$$\| x_n - x \| < \epsilon \quad \text{whenever } n \geq N(\epsilon)$$

in which case we write either $\| x_n - x \| \to 0$ as $n \to \infty$ or $x_n \to x$ as $n \to \infty$.

Definition 2.18

(i) The normed linear space $(X, \|\cdot\|)$ is **complete** if it is complete as a metric space in the induced metric.

(ii) A complete normed linear space is called a **Banach space**.

However, we would emphasise that there are metric spaces which are not normed linear spaces; a comparison of Definition 2.2 and Definition 2.14 clearly indicates this.

Definition 2.19

A **bounded linear operator** from a normed linear space $(X_1, \|\cdot\|_1)$ to a normed linear space $(X_2, \|\cdot\|_2)$ is a mapping from X_1 to X_2 which satisfies

(i) $T(\alpha x + \beta y) = \alpha Tx + \beta Ty$, for all $\alpha, \beta \in \mathbf{K}$, $x, y \in X_1$.

(ii) there exists a constant $c \geq 0$ such that $\| Tx \|_2 \leq c \| x \|_1$ for all $x \in X_1$. The smallest such c is called the **norm** of the operator T and is denoted $\| T \|$.

Consequently

$$\| T \| := \sup\left\{ \frac{\| Tx \|_2}{\| x \|_1} : \| x \|_1 \neq \theta \right\}.$$

The set of all bounded, linear operators from $(X_1, \|\cdot\|_1)$ to $(X_2, \|\cdot\|_2)$ will be denoted $B(X_1, X_2)$ or simply $B(X)$ when the two normed linear spaces are identical.

Notation 2.20

When we write $T : X_1 \to X_2$ then it is understood that $D(T)$, the **domain** of T, satisfies $D(T) = X_1$. The **range** of T, denoted $R(T)$, is defined to be

$$R(T) = \{y \in X_2 : y = Tx, \quad x \in D(T) \subseteq X_1\}.$$

The **null space** of T, denoted $N(T)$ is defined as

$$N(T) = \{x \in D(T) \subseteq X_1 : Tx = \theta \in X_2\}.$$

Definition 2.21

Let $T : X_1 \to X_2$ denote a linear operator between the two normed linear spaces $(X_k, \|\cdot\|_k)$, $k = 1, 2$.

(i) T is continuous at $x \in X_1$ if, whenever $\{x_n\}_{n=1}^{\infty} \subset X_1$ is a sequence such that

$$x_n \to x \quad \text{with respect to } \|\cdot\|_1$$

then

$$Tx_n \to Tx \quad \text{with respect to } \|\cdot\|_2 .$$

(ii) T is continuous on X_1 if it is continuous at all points $x \in X_1$.

The following results will be used frequently

Theorem 2.22

Let $T : X_1 \to X_2$ denote a linear operator between the two normed linear spaces $(x_k, \|\cdot\|_k)$, $k = 1, 2$. The following statements are equivalent.

(i) T is continuous at a point in X_1.

(ii) T is continuous on X_1.

(iii) T is bounded.

Theorem 2.23

Let $(X_k, \|\cdot\|_k)$, $k = 1, 2$ be complex Banach spaces

(i) $B(X_1, X_2)$ the set of bounded linear operators from X_1 into X_2 is a vector space with respect to the operations

$$\begin{aligned}(T_1 + T_2)x &= T_1 x + T_2 x, \quad T_1, T_2 \in B(X_1, X_2)\\ (\lambda T)x &= \lambda T x, \quad T \in B(X_1, X_2), \lambda \in \mathbf{C}.\end{aligned}$$

(ii) $B(X_1, X_2)$ is a complex Banach space with $\| T \|$ defined as in Definition 2.19.

(iii) $\| T \| = \sup \left\{ \frac{\|Tx\|_2}{\|x\|_1} : x \neq \theta \right\} = \sup \{\| Tx \|_2 : \| x \|_1 = 1\} \; \forall\, T \in B(X_1, X_2)$.

(iv) $\| Tx \|_2 \leq \| T \| \| x \|_1, \quad T \in B(X_1, X_2), \quad x \in X$.

2.3 Distribution Theory

The theory of distributions offers a generalisation of the concept of a function. It is a powerful mathematical tool for three reasons. First, in terms of distribution theory it is possible to give a precise description of idealised physical quantities such as point charges and instantaneous impulses. Second, it allows us to interchange limiting operations where such an interchange is not valid for classical functions. Consequently, in contrast to classical analysis, in distribution theory there are no problems arising from the existence of non–differentiable functions; that is, all distributions, or generalised functions as they are sometimes called, are infinitely differentiable. Third, it allows us to use series which in classical analysis we would call divergent.

Distribution theory arises as a result of the following observations. A continuous, complex valued function f of the n real variables $x = (x_1, \cdots, x_n)$ can be defined in two distinct ways. First we can prescribe its value $f(x)$ at each point x in $\mathbf{R}^n$. As we shall see, a second way of defining a continuous function f is to prescribe the value of the integral

$$I_f(\phi) = \int_{\mathbf{R}^n} f(x)\phi(x)dx \tag{2.1}$$

for each continuous, complex valued function ϕ whose value, $\phi(x)$, is zero for sufficiently large $| x |$. This integral always exists because $\phi(x)$ vanishes at infinity.

We shall need the following definitions

Definition 2.24

Elements of $B(X, \mathbf{K})$ the set of all bounded linear operators from the normed linear space X to the set of scalars $\mathbf{K}$ are called **linear functionals** on X. The space $B(X, \mathbf{K})$ is called the **dual space** of X and is denoted by X'.

For any $f \in X'$ we have, following Definition 2.19 since the norm on $\mathbf{K}$ is the modulus function,

$$\| f(x) \| = | f(x) | \leq c \| x \| \quad \text{for all } x \in X \text{ and some } c > 0$$

$$\| f \|_{X'} = \sup \left\{ \frac{| f(x) |}{\| x \|} : x \neq \theta \right\}.$$

Definition 2.25

Let a real or complex function f, defined on a domain $D \subset \mathbf{R}^n$, be non–zero only for points belonging to a subset $\Omega \subset D$. Then $\bar{\Omega}$, the closure of Ω, is called the **support** of f and is denoted by supp f.

A function f has compact support on D if its support is a closed, bounded subset of D.

Thus we see that in the second way of defining a function f we consider the functional I_f on the set of so-called '**test functions**' rather than the pointwise values $f(x)$.

The two descriptions of f are equivalent. To see this assume that for two continuous functions f and g the functionals I_f and I_g, defined as in (2.1) are equal; that is for any test function ϕ we have $I_f(\phi) = I_g(\phi)$. It then follows from elementary properties of the integral that $f(x) = g(x)$ for all $x \in \mathbf{R}^n$. The required equivalence is thus established.

A rationale for this second way of defining a function can be given as follows. A distributed physical quantity cannot be characterised by its value at a point but rather by its averaged value in a sufficiently close neighbourhood of that point. Consequently, from a physical standpoint it is more convenient to consider continuous functions as functionals.

In distribution theory we deal with functionals on an appropriate set of so-called test functions; the functionals are known as **distributions**. We would emphasise that in general a distribution does not have a definite value at a given point.

We now need some notation which can be conveniently introduced by means of an example.

Example 2.26

Let $C(\Omega)$ denote the set of all complex valued functions which are continuous on the region Ω. For $f \in C(\Omega)$ let

$$\| f \|_\infty := \sup\{| f(x) |: x \in \Omega\}.$$

It is readily verified that $C(\Omega)$ is a complex vector space with respect to the usual pointwise operations on functions. Furthermore, $C(\Omega)$ is a complete normed linear space with respect to $\| \cdot \|_\infty$.

Definition 2.27

(i) $C_0^\infty(\Omega) = \{\phi \in C^\infty(\Omega) : \operatorname{supp}\phi \subset \Omega\}$. In writing $\phi \in C^\infty(\Omega)$ we mean that ϕ and all the partial derivatives of ϕ of all orders exist and are continuous. An element $\phi \in C^\infty(\Omega)$ is also referred to as a smooth element.

(ii) The set $C_0^\infty(\Omega)$ is called the set of test functions.

(iii) A sequence of test functions $\{\phi_n\}_{n=1}^\infty$ is said to be convergent to a test function ϕ if

(a) ϕ and ϕ_n, $n = 1, 2, \cdots$ are defined on the same compact set, and

(b) for $x \in \mathbf{R}^n$

$$\| \phi_k(x) - \phi(x) \|_\infty \to 0 \text{ as } k \to \infty$$
$$\| D^\alpha \phi_k(x) - D^\alpha \phi(x) \|_\infty \to 0 \text{ as } k \to \infty$$

for all multi-indices α.

Note: The multi-index α and the mixed derivative D^α are introduced in Definition 2.34 below.

(iv) The set $C_0^\infty(\mathbf{R}^n)$ together with the topology induced by the convergence defined in (iii) is called the space of test functions and is denoted by $\mathcal{D}(\mathbf{R}^n)$.

(v) Any complex valued function f on $\mathcal{D}(\mathbf{R}^n)$ is called a functional; the value of f at a test function ϕ is denoted by $f(\phi)$.

A typical test function is the function ϕ defined by

$$\phi(x) = \begin{cases} 0, & | x | \geq a \\ \exp\{\frac{1}{x^2 - a^2}\}, & | x | < 1 \end{cases}$$

on $\Omega = (-b, b)$ where $b > a > 0$. Then $\phi \in C_0^\infty(\Omega)$ because it is readily shown that ϕ is infinitely, continuously differentiable and $\operatorname{supp}\phi = [-a, a]$.

Definition 2.28

(i) A **distribution** on a domain $\Omega \subset \mathbf{R}^n$ is a **continuous linear functional** on $\mathcal{D}(\mathbf{R}^n)$.

Thus a distribution f is a continuous linear mapping of the form

$$f : \mathcal{D}(\mathbf{R}^n) \to \mathbf{C}.$$

(ii) The space of distributions is the dual of $\mathcal{D}(\mathbf{R}^n)$, (see Definition 2.24) and is denoted by $\mathcal{D}(\mathbf{R}^n)'$.

Definition 2.29

(i) A function f which is integrable on every open, bounded subset $D \subset \Omega$ is called a **locally integrable** function on Ω.

(ii) A function f, locally integrable on Ω, defines a distribution $\tilde{f}$ on Ω according to

$$\tilde{f} : \mathcal{D}(\Omega) \to \mathbf{C}$$

$$\begin{aligned}\tilde{f}(\phi) &= \int_{\Omega} f(x)\phi(x)dx, \quad \phi \in \mathcal{D}(\Omega)\\ &=:< f, \phi >, \quad \phi \in \mathcal{D}(\Omega).\end{aligned}$$

The distribution $\tilde{f}$ is said to be generated by f.

(iii) Distributions of the type introduced in (ii) are called **regular distributions.** All other distributions are called **singular distributions.**

We would emphasise that for the definition given in (ii) to be meaningful we must have $B \equiv \operatorname{supp} \phi \subset \Omega$. In this case we have

$$\begin{aligned}| \tilde{f}(\phi) | &= \left| \int_{\Omega} f(x)\phi(x)dx \right| = \left| \int_{B} f(x)\phi(x)dx \right|\\ &\leq \sup_{x \in B} | \phi(x) | \int_{B} | f(x) | \, dx.\end{aligned}$$

The right hand side is bounded since f is locally integrable and we can conclude that $\tilde{f}(\phi)$ has meaning.

The use of different notations for a function f and its associated distribution $\tilde{f}$ will be supressed in future, whether f is a function or a distribution will be clear from the context.

Example 2.30

The Dirac delta, δ, defined by

(i) $\delta(x) = 0, \ x \neq 0$

(ii) $\int_{-\infty}^{\infty} \delta(x)dx = 1$

(iii) $\int_{-\infty}^{\infty} \delta(x)\phi(x)dx) = < \delta, \phi > = \delta(\phi) = \phi(0), \phi \in C_0^\infty(\mathbf{R}^n)$

is a continuous linear functional on $C[a, b]$, $b > a > 0$. That it is a linear functional is clear from (iii) whilst the continuity of δ follows from

$$| \delta(\phi) | = | < \delta, \phi > | = | \phi(0) | \leq \sup | \phi(x) | = \| \phi \|_\infty .$$

Every continuous function is locally integrable and hence generates a distribution (Definition 2.29(ii)) but there are also many irregular and discontinuous functions which are also locally integrable.

Example 2.31

$$f(x) = | x |^{-\frac{1}{2}}, \quad x \in [-1, 1].$$

This function has a singularity at the origin. However it is locally integrable since

$$\int_a^b | f(x) | \, dx = \int_a^b | x |^{-\frac{1}{2}} \, dx.$$

is bounded for each interval (a, b) in $[-1, 1]$. Thus f generates a distribution, which is also denoted by f, defined by

$$f(\phi) = < f, \phi > = \int_{-1}^{1} | x |^{-\frac{1}{2}} \phi(x)dx, \quad \phi \in C_0^\infty[-1, 1].$$

Thus f is a regular distribution.

Example 2.32

The Heaviside step function H defined on $[-a, a]$ by

$$H(x) = \begin{cases} 0, & -a \leq x < 0 \\ 1, & 0 \leq x \leq a \end{cases}$$

is locally integrable and generates the distribution H which satisfies

$$G(\phi) = < H, \phi > = \int_{-a}^{a} H(x)\phi(x)dx = \int_0^a \phi(x)dx \quad \phi \in C_0^\infty(-a, a).$$

Thus H is a regular distribution.

Example 2.33

The Dirac delta, δ, defined in Example 2.30 is a singular distribution. To show this let ϕ be a test function defined by

$$\phi(x) = \begin{cases} \exp\{\frac{a^2}{x^2-a^2}\} & , \mid x \mid < a \\ 0, & , b > \mid x \mid \geq a \end{cases}$$

where $b > a > 0$. Assume that δ is a regular distribution and show that

$$\mid \int_{-b}^{b} \delta(x)\phi(x)dx \mid \leq \frac{1}{e} \int_{-a}^{a} \delta(x)dx.$$

Then by considering the limit as $a \to 0$ obtain a contradiction.

The Dirac delta and Heaviside step function do not have derivatives in the classical sense. However, it is possible to extend the classical notion of derivative in such a way that a derivative for these quantities and indeed for any distribution may be defined. Furthermore, should the distribution be a continuously differentiable function then the extended notion of a derivative collapses to the ordinary definition of a derivative.

In order to simplify the notation we introduce the so–called multi-indices.

Definition 2.34

(i) A set $\alpha = (\alpha_1, \alpha_2, \cdots, \alpha_n)$ of n non-negative integers is called a **multi-index**.

(ii) $\mid \alpha \mid = \sum_{k=1}^{n} \alpha_k$ is the **order** of the multi-index.

(iii) We denote

(a) $f_\alpha(x) = f_{\alpha_1 \alpha_2 \cdots \alpha_n}(x)$

(b) $D^\alpha f(x) = D_1^{\alpha_1} D_2^{\alpha_2} \cdots D_n^{\alpha_n} f(x)$, where $D_k^m = \dfrac{\partial^m}{\partial x_k^m}$.

The starting point for obtaining an extended definition of a derivative is the familiar integration by parts formula

$$\int_a^b u(x)\frac{dv}{dx}(x)dx = [u(x)v(x)]_a^b - \int_a^b v(x)\frac{du}{dx}(x)dx. \tag{2.2}$$

This result is easily generalised, using Green's Theorem, to one involving partial derivatives of order m. Specifically we can obtain for $u, v \in C^m(\bar{\Omega})$,

$$\int_\Omega (D^\alpha u)(x)v(x)dx = (-1)^{|\alpha|} \int_\Omega u(x)D^\alpha v(x)dx + \int_\Gamma H(u,v)ds \tag{2.3}$$

where $\Omega \subset \mathbf{R}^n$ is a region with boundary Γ and $H(u,v)$ is an expression involving a sum of products of derivatives of u and v of order less than m.

If we replace v in (2.3) by a test function $\phi \in C_0^\infty(\Omega)$ then, since ϕ and its derivatives of all orders vanish on the boundary, (2.3) reduces to

$$\int_\Omega (D^\alpha u)(x)\phi(x)dx = (-1)^{|\alpha|} \int_\Omega u(x)D^\alpha \phi(x)dx. \tag{2.4}$$

Since $u \in C^m(\bar{\Omega})$ then it generates a regular distribution, also denoted by u so that

$$u(\phi) = < u, \phi > = \int_\Omega u(x)\phi(x)dx. \tag{2.5}$$

Furthermore, since $\phi \in C_0^\infty(\Omega)$ then we also have that $D^\alpha \phi \in C_0^\infty(\Omega)$. Consequently $D^\alpha \phi$ is a test function so that $u \in C^m(\bar{\Omega})$ generates the distribution

$$u(D^\alpha \phi) = < u, D^\alpha \phi > = \int_\Omega u(x)D^\alpha \phi(x)dx. \tag{2.6}$$

On the other hand $D^\alpha u$ is continuous and as such is able to generate a distribution, also denoted by $D^\alpha u$, satisfying

$$D^\alpha u(\phi) = < D^\alpha u, \phi > = \int_\Omega D^\alpha u(x)\phi(x)dx. \tag{2.7}$$

Therefore, using (2.6) and (2.7) we see that (2.4) can be written

$$D^\alpha u(\phi) = (-1)^{|\alpha|} u(D^\alpha \phi). \tag{2.8}$$

We take (2.8) to be the basis for defining the derivative of order $|\alpha|$ of any distribution.

Definition 2.35

The α-th **distributional** or **generalised derivative** of a distribution f on Ω is a distribution, denoted by $D^\alpha f$, which satisfies

$$D^\alpha f(\phi) = (-1)^{|\alpha|} f(D^\alpha \phi) \quad \text{for all } \phi \in C_0^\infty(\Omega). \tag{2.9}$$

Example 2.36

The Heaviside step function H has a distributional derivative, denoted by H' satisfying

$$\begin{aligned} H'(\phi) &= < H', \phi > = (-1)H(\frac{d\phi}{dx}), \quad \phi \in C_0^\infty(-1,1) \\ &= -\int_{-1}^{1} H(x)\frac{d\phi}{dx}(x)dx = -\int_{0}^{1} \frac{d\phi}{dx}(x)dx = [\phi(x)]_0^1 \\ &= \phi(0) = \delta(\phi) \end{aligned}$$

Thus $H' = \delta$ in the distributional sense.

The distributional derivative, $D^\alpha f$, of a regular distribution f on Ω is defined by (2.9). However, in general it cannot be assumed that $D^\alpha f$ is also a regular distribution. In the event that $D^\alpha f$ is a regular distribution, then, as we have seen, it is generated by the locally integrable function $D^\alpha f$ according to

$$D^\alpha f(\phi) = \int_\Omega D^\alpha f(x)\phi(x)dx, \quad \phi \in C_0^\infty(\Omega). \tag{2.10}$$

Consequently, combining (2.9) and (2.10) we see that u and $D^\alpha u$ are related according to

$$\int_\Omega D^\alpha f(x)\phi(x)dx = (-1)^m \int_\Omega f(x)D^\alpha \phi(x)dx \tag{2.11}$$

where $\mid \alpha \mid = m$, and $\phi \in C_0^\infty(\Omega)$.

We use (2.11) as a general definition of the quantity $D^\alpha u$, and shall refer to $D^\alpha u$ as the **weak derivative** of u. We notice that whenever u is sufficiently smooth, for example $u \in C^m(\Omega)$, then $D^\alpha u$, the weak derivatives of u, coincides with the **classical derivatives** of u for $\mid \alpha \mid \leq m$.

Example 2.37

The function $f \in C(-1,1)$ defined by $f(x) = \mid x \mid$, $x \in (-1,1)$ does not have a classical derivative at $x = 0$. However, since for all $\phi \in C_0^\infty(-1,1)$

$$f'(\phi) = (-1)f(\phi') = (-1)\int_{-1}^{1} f(x)\phi'(x)dx = \int_{-1}^{1} (2H(x) - 1)\phi(x)dx$$

where H is the Heaviside function (Example 2.32) then we see from (2.11) that f has a weak derivative

$$f'(x) = 2H(x) = 1.$$

We notice that this derivative exists, in a distributional sense, at the origin and furthermore that f' is locally integrable.

We notice that we have used the expression $D^\alpha f$ to denote the α–th partial derivative of f in the sense of either a classical derivative or a weak derivative as defined in (2.11) or a distributional (generalised) derivative as defined in (2.9). This will not cause confusion as it will always be clear from the text which derivative is being used.

It is now natural to investigate differential equations and their solutions from the standpoint of distributions. Before doing this we need certain **local properties** of distributions.

If f is a distribution then it does not make any sense to say "f vanishes at a point x". However, the local statment " f vanishes in an open region Ω" does make sense. This is because we interpret the last statement as follows: $f = 0$ in Ω if and only if $< f, \phi >$ for all test functions ϕ such that supp $\phi \subset \Omega$. A regular distribution f has the property $f = 0$ in Ω if and only if $f(x) = 0$ for all $x \in \Omega$.

The product gf of a smooth function g which vanishes in some open region Ω and an arbitrary distribution f always vanishes in Ω. Furthermore, if $f = 0$ in Ω then $D^\alpha f = 0$ in Ω for every multi-index α.

We say that the minimal closed set F such that $f = 0$ in $\mathbf{R}^n \backslash F$ is the **support** of the distribution f and it is denoted by $\operatorname{supp} f$. Clearly, $< f, \phi >= 0$ if $\operatorname{supp} f$ and $\operatorname{supp} \phi$ do not intersect. A distribution f is said to be **finite** or of **compact support** if $\operatorname{supp} f$ is compact. Finally, we say that a distribution f is **localised** at a point x if $\operatorname{supp} f = x$.

With this preparation we now turn our attention to equations of the form

$$Lu = f \tag{2.12}$$

where L is a differential expression of the form

$$L = a_o(x)\frac{d^k}{dx^k} + a_1(x)\frac{d^{k-1}}{dx^{k-1}} + \cdots + a_k(x). \tag{2.13}$$

If f is a continuous function and L contains only classical derivatives and sufficiently smooth functions a_m, $m = 0, 1, \cdots, k$, then a solution u of (2.12) should be a k times continuously differentiable function. This is indeed the case and u is known as a **classical solution** of (2.12).

Suppose now that we require to find a distribution u which satisfies (2.12). In this case the differential equation (2.12) is interpreted as a differential equation involving distributional derivatives and we look for solutions u which satisfy

$$Lu(\phi) = f(\phi), \quad \phi \in C_0^\infty(\Omega) \tag{2.14}$$

where Ω denotes the region over which the differential equation is defined. The equation (2.14) is equivalent to

$$u(L^*\phi) = f(\phi), \quad \phi \in C_0^\infty(\Omega) \tag{2.15}$$

where L^* is the differential expression which arises as a consequence of applying (2.9) successively. Specifically

$$L^*\phi(x) = (-1)^k \frac{d^k}{dx^k}(a_o(x)\phi(x)) + \cdots + a_k(x)\phi(x). \tag{2.16}$$

If f is a regular distribution generated by a function which is locally integrable but not continuous then the equation (2.12) cannot be expected to have any meaning in a classical sense. A similar observation holds if f is a singular distribution. The solutions of (2.12) in these cases, that is distributions u which satisfy (2.14), are known as **weak** or **generalised solutions** and we say that (2.12) holds in the **sense of distributions.**

Of course, whenever the distributions involved in a discussion of (2.12) are generated by sufficiently differentiable functions then we recover the classical concepts of a differential equation and its solutions.

We would remark that although the above discussion has been conducted with respect to an ordinary differential equation nevertheless exactly the same remarks can be made for partial differential equations.

Example 2.38

Consider the equation

$$xu'(x) = 0 \quad x \in \Omega = (-1, 1). \tag{2.17}$$

This has a classical solution

$$u(x) = \text{ constant.}$$

However, regarding (2.17) as a distributional differential equation it has a weak solution

$$u(x) = c_1 H(x) + c_2$$

where c_1, c_2 are constants. This can be seen to be the case if we first notice (Example 2.36) that

$$u'(x) = c_1 \delta(x)$$

so that

$$(xu')(\phi) = u'(x\phi) = c_1 \delta(x\phi) = c_1 (x\phi)(0) = 0$$

and we conclude that $xu' = 0$ in the sense of distributions.

The above example is particularly instructive as it involves the product of a distribution and a continuous function. Such a product can be defined in a natural way.

Definition 2.39

If u is a distribution and f a continuous function on $\Omega \subset \mathbf{R}^n$ then the **product** fu is interpreted as the distribution which satisfies

$$fu(\phi) = u(f\phi) \quad \phi \in C_0^\infty(\Omega).$$

We notice that our definition is a generalisation of the familiar identity

$$\int_\Omega \{f(x)u(x)\}\phi(x)dx = \int_\Omega u(x)\{f(x)\phi(x)\}dx$$

which holds when u is locally integrable.

Example 2.40

The distribution $f\delta$ on (-1,1) with $f(x) = x$ satisfies

$$x\delta(\phi) = \delta(x\phi) = [x\phi]_{x=0} = 0$$

a result which holds for all $\phi \in C_0^\infty(-1,1)$. Thus, $x\delta =: \Theta \subset (C_0^\infty(-1,1))'$ and is the zero distribution.

The classical formula for the **convolution** of two continuous functions f and g is given by

$$(f * g)(x) = \int_{\mathbf{R}} f(x-y)g(y)dy. \tag{2.18}$$

When f and g are distributions we consider the relation

$$< (f * g)(x), \phi(x) >=< f(x), < g(y), \phi(x+y) >> \tag{2.19}$$

where we have abused notation slightly in order to emphasise the 'integration' variable. In (2.19) we have assumed that g has compact support and that ϕ is a test function. Consequently the right hand side of (2.19) is well defined since

$$\psi(x) := < g(y), \phi(x+y) >$$

is a test function. We notice that when f and g are regular distributions then $f * g$ is also a regular distribution and (2.19) implies the classical formula (2.18).

Example 2.41

The Dirac delta concentrated at $x = a$, sometimes denoted by $\delta_a(x)$, is defined by

$$\delta_a(x) = \delta(x-a)$$

and has the property (see Definition 2.30)

$$< \delta_a(x), \phi(x) > = \int_{\mathbf{R}} \delta_a(x)\phi(x)dx = \phi(a).$$

The convolution of a regular distribution f with δ_a is, by (2.19),

$$\begin{aligned} < (f * \delta_a)(x), \phi(x) > &= < f(x), < \delta_a(y), \phi(x+y) >> \\ &= < f(x), \phi(x+a) > \\ &= < f(x-a), \phi(x) > \end{aligned}$$

Thus, remembering that the 'integration' variables are only written in for convenience we see that

$$f(x) * \delta_a(x) = f(x-a)$$

in a distributional sense.

The Fourier transform of distributions will play an important rôle in later sections. We shall not define the Fourier transform of arbitrary distributions but rather confine attention to so-called **tempered distributions**. With this in mind we shall need the following

Definition 2.42

A smooth function ϕ is said to be a **rapidly decreasing** function if it satisfies inequalities of the form

$$(1+ \mid x \mid^m) \mid D^\alpha \phi(x) \mid \leq c(m, \alpha) \qquad (2.20)$$

where $m \geq 0$ is an index, α is a multi–index with $| \alpha | \geq 0$ and $c(m, \alpha)$ is an appropriate constant.

We notice that any smooth function with compact support satisfies (2.20) as also does the function ϕ defined by

$$\phi(x) = \exp(- | x |^2).$$

It can be shown that the set of rapidly decreasing functions is, under quite natural laws of composition, a vector space. We would remark that the product of a rapidly decreasing function and a continuous function is not, in general, a rapidly decreasing function.

Convergence in the set of rapidly decreasing functions is defined according to

Definition 2.43

A sequence of rapidly decreasing functions $\{\phi_n\}_{n=1}^{\infty}$ is said to be convergent to a rapidly decreasing function ϕ if for all m and α the numerical sequence defined by

$$\sup_x \{(1+ | x |^m) | D^\alpha(\phi_n - \phi) |\}$$

is convergent to zero.

Notation 2.44

We denote by $S(\mathbf{R}^n)$ the space of rapidly decreasing functions endowed with the convergence defined in Definition 2.43.

We would emphasise that $(D_i\phi)(x)$ and $x_i\phi(x)$ belong to $S(\mathbf{R}^n)$ whenever $\phi \in S(\mathbf{R}^n)$.

It is clear that a sequence of test functions $\{\phi\}_{n=1}^{\infty} \subset \mathcal{D}(\mathbf{R}^n)$ which is convergent in $\mathcal{D}(\mathbf{R}^n)$ is also convergent in $S(\mathbf{R}^n)$. Consequently any linear, continuous functional on $S(\mathbf{R}^n)$ is a distribution; we shall refer to such distributions as **tempered distributions.** These distributions form a linear space endowed with the standard convergence condition given in Definition 2.27; this space is denoted by $S'(\mathbf{R}^n)$. We would emphasise that $S'(\mathbf{R}^n)$ is closed with respect to differentiation D_i and multiplication by x_i since $S(\mathbf{R}^n)$ is closed with respect to these operations.

The **Fourier transform** of $\phi \in S(\mathbf{R}^n)$ has the form

$$\hat{\phi}(k) = (F\phi)(k) = \frac{1}{(2\pi)^{n/2}} \int_{\mathbf{R}^n} e^{-ik.x} \phi(x) dx \tag{2.21}$$

where $x, k \in \mathbf{R}^n$ and $k.x = \sum_{i=1}^{n} k_i x_i$.

The inverse Fourier transform has the similar form

$$\psi(x) = (F^{-1}\hat{\psi})(x) = \frac{1}{(2\pi)^{n/2}} \int_{\mathbf{R}^n} e^{ik.x} \hat{\psi}(k) dk. \tag{2.22}$$

It can be shown that

(i) $\hat{\phi}$ is a rapidly decreasing function.

(ii) $F : S(\mathbf{R}^n) \to S(\mathbf{R}^n)$, is linear and one-one.

(iii) Parseval's equality holds in the form

$$\int_{\mathbf{R}^n} \phi(x) \hat{\psi}(x) dx = \int_{\mathbf{R}^n} \hat{\phi}(k) \psi(k) dk.$$

We define the Fourier transform of a tempered distribution as follows.

Definition 2.45

The Fourier transform of a tempered distribution f is $\hat{f}(k) = (Ff)(k)$ where

$$< \hat{f}, \phi > = < f, \hat{\phi} >, \quad \phi \in S(\mathbf{R}^n).$$

It can be shown that $\hat{f}$ is a tempered distribution.

Furthermore, the Fourier transform, F, is a linear, one–one mapping of $S'(\mathbf{R}^n)$ into itself.

To define the inverse Fourier transform, F^{-1}, of a distribution $f(k)$ it is convenient to re-write (2.22) in the more suggestive form

$$\check{\psi}(x) = (F^{-1}\psi)(x) = \frac{1}{(2\pi)^{n/2}} \int e^{ik.x} \psi(k) dk. \tag{2.22(a)}$$

With this notation the required inverse is denoted by

$$\check{f}(x) = (F^{-1}f)(x) \tag{2.23}$$

where

$$< \check{f}, \phi > = < f, \check{\phi} >, \quad \phi \in S(\mathbf{R}^n). \tag{2.24}$$

The Fourier transforms of a tempered distribution f and its distributional derivative $D_i f$, which is also a tempered distribution, are connected by the formulae

$$\begin{aligned} F((D_i f))(k) &= ik_i \hat{f}(k), \quad D_i = \partial/\partial x_i \\ F(ix_i f)(k) &= -D_i \hat{f}(k), \quad D_i = \partial/\partial k_i \end{aligned} \tag{2.25}$$

where $x = (x_1, \cdots, x_n)$ and $k = (k_1, \cdots, k_n)$ are elements of $\mathbf{R}^n$.

In general, if P is a polynomial then it is a straightforward matter to show that

$$\begin{aligned} F((P(D)f)(k) &= P(ik)\hat{f}(k), \quad D = \partial/\partial x \\ F((P(ix))f)(k) &= P(-D)\hat{f}(k), \quad D = \partial/\partial k. \end{aligned} \tag{2.26}$$

Example 2.46

$$\hat{\delta}(k) = (F(\delta))(k) = (F\delta)(k)$$

$$\begin{aligned} < \hat{\delta}, \phi > = < \delta, \hat{\phi} > = \hat{\phi}(0) &= \frac{1}{(2\pi)^{n/2}} \left\{ \int_{\mathbf{R}^n} e^{ik \cdot x} \phi(x) dx \right\}_{k=0} \\ &= \frac{1}{(2\pi)^{n/2}} < 1, \phi > . \end{aligned}$$

Thus

$$\hat{\delta}(k) = \frac{1}{(2\pi)^{n/2}}.$$

Finally, it can be shown

(i) $\hat{1} = (2\pi)^{-n/2}\delta(k), \quad k \in \mathbf{R}^n.$

(ii) $(F(P(D)\delta))(k) = (2\pi)^{-n/2}P(ik), \quad D = \partial/\partial x, \quad x \in \mathbf{R}^n.$

$F(P(ix))(k) = (2\pi)^{-n/2}P(-D)\delta(k), \quad D = \partial/\partial k, \quad x \in \mathbf{R}^n.$

(iii) $(F(1/x))(k) = -(\pi/2)^{1/2} i sgn k, \quad x \in \mathbf{R}.$

The details are left as an exercise. However, for $x = (x_1, x_2, ... x_n) \in \mathbf{R}^n$ we understand

$$\delta(x) = \delta(x_1)\delta(x_2)...\delta(x_n).$$

2.4 Hilbert space

A frequently occurring type of Banach space is the one whose norm can be defined in terms of an **inner product**, a generalisation of the familiar dot product on finite dimensional vector spaces. These spaces have quite a rich structure much of which depends on the notion of angle which is implicit in their definition.

Definition 2.47

A complex vector space X is called an **inner product space** if there is defined on $X \times X$ a complex valued function $(\cdot, \cdot)$, called an **inner product**, which satisfies for all $x, y \in X$ and $\alpha \in \mathbf{C}$

(i) $(x, y) \geq 0$.

(ii) $(x, x) = 0$ if and only if $x = \theta$.

(iii) $(\alpha x, y) = \alpha(x, y), \quad (x, \alpha y) = \bar{\alpha}(x, y)$.

(iv) $(x + y, z) = (x, z) + (y, z)$.

(v) $(x, y) = \overline{(y, x)}$.

where the bar denotes complex conjugate.

Example 2.48

(i) On $\mathbf{C}^n$, the set of n–tuples of complex numbers $x = (x_i, \cdots, x_n)$, $y = (y_1, \cdots, y_n)$, we can define

$$(x, y) = \sum_{k=1}^{n} x_k \bar{y}_k.$$

(ii) On $C[a, b]$, the set of complex valued continuous functions defined on the interval $[a, b]$, we can define

$$(f, g) = \int_a^b f(x)\overline{g(x)}dx, \quad f, g \in C[a, b].$$

Definition 2.49

Let X be an inner product space.

(i) Elements $x, y \in X$ are said to be **orthogonal** if $(x, y) = 0$.

(ii) A set of elements $\{x_k\}_{k=1}^{\infty} \subset X$ is called **orthonormal** if

$$(x_k, x_j) = \delta_{kj} = \begin{cases} 1, & k = j \\ 0, & k \neq j. \end{cases}$$

We shall now state a number of fundamental results of Hilbert space theory the proofs of which can be found in the references.

Theorem 2.50

Every inner product space X is a normed linear space with norm defined by

$$\| x \| = (x, x)^{1/2}, \quad x \in X. \tag{2.27}$$

Theorem 2.51 (Theorem of Pythagoras)

If $\{x_k\}_{k=1}^{N}$ is an orthonormal set in an inner product space X then for all $x \in X$

$$\| x \|^2 = \sum_{k=1}^{N} | (x, x_k) |^2 + \| x - \sum_{k=1}^{N} (x, x_k) x_k \|^2 .$$

Corollary 2.51.1: (Bessel's inequality)

$$\| x \|^2 \geq \sum_{k=1}^{N} | (x, x_k) |^2 . \tag{2.28}$$

Corollary 2.51.2: (Schwarz inequality)

If x, y are any two elements in an inner product space X then

$$| (x, y) | \leq \| x \| \, \| y \| . \tag{2.29}$$

These various results are readily seen to be generalisations of familiar results in finite dimensional vector analysis. A further result which will prove to be useful is the **parallelogram law**; if x, y are any two elements in an inner product space X then

$$\| x + y \|^2 + \| x - y \|^2 = 2 \| x \|^2 + 2 \| y \|^2 . \tag{2.30}$$

Definition 2.52

(i) An inner product space is called a pre-Hilbert space.

(ii) A complete inner product space is called a **Hilbert** space.

Definition 2.53

Two Hilbert spaces H_1, H_2 are **isomorphic** if there is a linear operator $T : H_1 \to H_2$ (onto) such that

$$(Tx, Ty)_2 = (x, y)_1 \quad \text{for all } x, y \in H_1.$$

Here $(\cdot, \cdot)_k$ denotes the inner product in H_k, $k = 1, 2$. The operator T is called a **unitary** operator.

Example 2.54

(i) Let H_k be a Hilbert space with inner product $(\cdot, \cdot)_k$, $k = 1, 2$. The set of ordered pairs of the form

$$< x, y >, \quad x \in H_1, \quad y \in H_2$$

is a Hilbert space H with inner product

$$(< x_1, y_1 >, < x_2, y_2 >) := (x_1, x_2)_1 + (y_1, y_2)_2.$$

H is called the direct product of H_1 and H_2 and is denoted by $H = H_1 \times H_2$.

(ii) Let $\{H_k\}_{k=1}^{\infty}$ be a sequence of Hilbert spaces with structure $(\cdot, \cdot)_k$, $\| \cdot \|_k$, $k = 1, 2, \cdots$ Let H denote the set of sequences $\{x_k\}_{k=1}^{\infty}$, $x_k \in H_k$ which satisfy $\sum_{k=1}^{\infty} \| x_k \|_k^2 < \infty$. Then H is a Hilbert space with inner product

$$(x, y) = \sum_{k=1}^{\infty} (x_k, y_k), \quad x, y \in H$$

and we write $H = \mathop{\times}\limits_{k=1}^{\infty} H_k$.

One of the main reasons why Hilbert spaces are easier to handle than Banach spaces is indicated by the so-called Projection Theorem. However, before stating this important result we would emphasise the following concepts.

A subset M of a vector space V is called a **linear manifold** if for all $x, y \in V$ and $\alpha, \beta \in \mathbf{K}$ we have $(\alpha x + \beta y) \in M$.

If V is a normed linear space and if M is closed in the structure of V then the **closed linear manifold** M is referred to as a **subspace** of V.

Theorem 2.55: (Projection Theorem)

Let

(i) H be a Hilbert space with inner product $(\cdot, \cdot)$.

(ii) M be a subspace of H.

(iii) $M^{\perp}$ be the set of elements in H which are orthogonal to M.

Then every element $x \in H$ can be written uniquely in the form

$$x = y + z, \quad y \in M, z \in M^{\perp}.$$

This we denote by writing

$$H = M \oplus M^{\perp} = \{y + z : y \in M, z \in M^{\perp}\}$$

and we refer to H as the **direct sum** of M and $M^{\perp}$. The word "direct" is used to emphasise that $M \cap M^{\perp} = \{0\}$.

Since all Hilbert spaces are Banach spaces much of the notation already introduced carries over to this section. In particular $B(H_1, H_2)$ denotes the set of all bounded linear operators from H_1 to H_2 and as before $B(H_1, H_2)$ is a Banach space with respect to the norm

$$\| T \| = \sup\{\| Tx \|_2 : \| x \|_1 = 1\}, \quad T \in B(H_1, H_2). \tag{2.31}$$

Definition 2.56

The space $B(H, \mathbf{K})$ is called the **dual space** of the Hilbert space H and is denoted by H^*. The elements of H^* are continuous linear functionals on H.

The next result has considerable importance in applications as it characterises the elements of H^*.

Theorem 2.57 (Riesz Representation Theorem)

For each element $f \in H^*$ there exists an element $y \in H$ such that

$$f(x) = (x, y)_H \quad \text{for all } x \in H.$$

The element y clearly depends on f. Furthermore

$$\| f \|_{H^*} = \| y \|_H .$$

There are two notions of convergence in Hilbert spaces.

Definition 2.58

Let H be a Hilbert space with structure $(\cdot, \cdot)$, $\| \cdot \|$.

(i) A sequence $\{x_k\}_{k=1}^{\infty} \subset H$ is said to **converge strongly** to an element $x \in H$ if

$$\lim_{k \to \infty} \| x_k - x \| = 0$$

in which case we write

$$s - \lim_{k \to \infty} x_k = x.$$

(ii) A sequence $\{x_k\}_{k=1}^{\infty} \subset H$ is said to **converge weakly** to an element $x \in H$ if

$$\lim_{k \to \infty} (x_n, y) = (x, y) \quad \text{for all } y \in H$$

in which case we write

$$w - \lim_{k \to \infty} x_k = x.$$

Theorem 2.59

Let $\{x_k\}_{k=1}^{\infty} \subset H$ and $x \in H$. Then $s - \lim_{k \to \infty} x_k = x$ if and only if

(i) $w - \lim_{k \to \infty} x_k = x$

and

(ii) $\lim_{k \to \infty} \| x_k \| = \| x \|$.

The next results enable us to characterise elements of a Hilbert space.

Definition 2.60

If X is an orthonormal set in a Hilbert space H and there is no other orthonormal set in H which contains X as a proper subset then X is called an **orthonormal basis or complete orthonormal system for** H.

Theorem 2.61

(i) Every Hilbert space has an orthonormal basis.

(ii) Let H be a Hilbert space and $X = \{x_\alpha\}_{\alpha \in I}$ an orthonormal basis for H. For each $y \in H$

$$y = \sum_{\alpha \in I} (y, x_\alpha) x_\alpha \tag{2.32}$$

$$\| y \|^2 = \sum_{\alpha \in I} | (y, x_\alpha) |^2 \tag{2.33}$$

where I is an arbitrary index set. However, we shall be mainly concerned with the case when I is a set of integers.

The coefficients (y, x_α) of x_α in (2.32) are called the Fourier coefficients of y with respect to the basis $\{x_\alpha\}_{\alpha \in I}$. The result (2.33) is known as **Parseval's relation.**

We would remark that, unless there are statements to the contrary, expressions such as (2.32) will always be understood to imply convergence with respect to a strong limit, that is

$$\sum_{k=1}^{\infty} \alpha_k x_k = s - \lim_{N \to \infty} \sum_{k=1}^{N} \alpha_k x_k. \tag{2.34}$$

Example 2.62

$y \in H$ implies $\| y \| < \infty$ and from (2.33) since the series must be convergent it follows that

$$\lim_{k \to \infty} (x_k, y) = 0 = (0, y)$$

the last equality holding because y is arbitrary. Thus we see that the basis elements x_k converge weakly to zero. However they do not converge strongly since

$$\lim_{k \to \infty} \| x_k \| = \lim_{k \to \infty} 1 = 1 \neq 0.$$

Definition 2.63

A metric space, and hence a Hilbert space, is said to be **separable** if it has a countable, dense subset.

Theorem 2.64

A Hilbert space H is separable if and only if it has a countable orthonormal basis.

One of the simplest but perhaps most important Hilbert spaces is $L_2(\Omega)$, the space of square integrable functions defined on the region $\Omega \subseteq \mathbf{R}^n$. An inner product for $L_2(\Omega)$ may be defined by

$$(f,g)_o = \int_\Omega f(x)\overline{g(x)}dx.$$

The associated norm is denoted by $\|\cdot\|_o$.

In order to be able to account for the differentiability properties of elements in $L_2(\Omega)$ we shall work in the framework of so-called Sobolev spaces. Of particular relevance to us here will be the Sobolev-Hilbert spaces, denoted by $H^m(\Omega)$, consisting of those elements defined on Ω which, together with all their weak derivatives up to and including those of order m, belong to $L_2(\Omega)$:

$$H^m(\Omega) := \{f : D^\alpha f \in L_2(\Omega), \quad |\alpha| \leq m\}.$$

An inner product on $H^m(\Omega)$ may be defined by

$$(f,g)_m = \int_\Omega \sum_{|\alpha|\leq m} D^\alpha f(x)\overline{D^\alpha g(x)}dx.$$

We notice that $H^o(\Omega) = L_2(\Omega)$ and that

$$\| f \|_m^2 = \sum_{|\alpha|\leq m} \| D^\alpha u \|_o^2 .$$

2.5 Bounded linear operators on Hilbert spaces.

We have seen that the concept of a bounded linear operator can be interpreted in a Hilbert space setting. In this section we collect the more frequently used results for such operators. In this connection the reader should refer to the material from Definition 2.14 to the end of §2.2 and give it an interpretation in a Hilbert space setting.

In dealing with bounded linear operators it is convenient to introduce the following notions of convergence.

Definition 2.65

Let H be a Hilbert space, $\{T_k\}_{k=1}^{\infty} \subset B(H)$ and $T \in B(H)$. We say that

(i) $\{T_k\}_{k=1}^{\infty}$ is **strongly convergent** to T if

$$\lim_{k\to\infty} \| T_k x - Tx \| = 0 \quad \text{for all } x \in H$$

and we write $s-\lim_{k\to\infty} T_k = T$.

(ii) $\{T_k\}_{k=1}^{\infty}$ is **weakly convergent** to T if $\{T_k x\}$ is weakly convergent to Tx for each $x \in H$ and we write $w - \lim_{k\to\infty} T_k = T$.

(iii) $\{T_k\}_{k=1}^{\infty}$ is **uniformly convergent** to T if

$$\lim_{k\to\infty} \| T_k - T \| = 0$$

and we write $u - \lim_{k\to\infty} T_k = T$.

Uniform convergence is sometimes called convergence in norm.

Theorem 2.66

Let H be a Hilbert space and $T_k, T, S_k, S \in B(H)$, $\quad k = 1, 2, \cdots$.

(i) If $s-\lim_{k\to\infty} T_k = T$ and $s-\lim_{k\to\infty} S_k = S$ then $s - \lim_{k\to\infty} T_k S_k = TS$.

(ii) If $u-\lim_{k\to\infty} T_k = T$ and $u-\lim_{k\to\infty} S_k = S$ then $u - \lim_{k\to\infty} T_k S_k = TS$.

We would emphasise that a similar result for weak convergence does **not** hold.

Definition 2.67

A bounded linear operator T on H is **invertible** if there exists a bounded linear operator T^{-1} on H such that

$$TT^{-1} = T^{-1}T = I$$

where I is the **identity** operator on H.

Theorem 2.68

An invertible operator $T \in B(H)$ is one-one and maps H onto H. The inverse of T is unique.

The next result, which provides a generalisation of the geometric series for $(1-a)^{-1}$, is fundamental to much of the material in later sections.

Theorem 2.69 (Neumann series)

Let H be a Hilbert space and $T \in B(H)$ with the property that $\| T \| < 1$. Then

(i) $(I-T)$ is invertible.

(ii) $(I-T)^{-1} \in B(H)$.

(iii) $(I-T)^{-1} = \sum_{k=0}^{\infty} T^k$ (Neumann series)

is uniformly convergent.

(iv) $\| (I-T)^{-1} \| \leq (I - \| T \|)^{-1}$

Proof We notice first that if $T_k \in B(H)$, $k = 1,2$ then from the definition of $\| T_k \|$ and the triangle inequality we obtain

$$\begin{aligned} \| T_1 + T_2 \| &\leq \| T_1 \| + \| T_2 \| \\ \| T_1 T_2 \| &\leq \| T_1 \| \, \| T_2 \| . \end{aligned} \tag{2.35}$$

Therefore

$$\| T^k \| = \| T T^{k-1} \| \leq \| T \| \| T^{k-1} \| \leq \cdots \leq \| T \|^k .$$

Since $\| T \| < 1$ we conclude that $\| T^k \| \to 0$ as $k \to \infty$ that is

$$u - \lim_{k \to \infty} T^k = 0. \tag{2.36}$$

From the triangle inequality we have

$$\begin{aligned} \| \sum_{k=n}^{m} T^k \| &\leq \sum_{k=n}^{m} \| T \|^k = \| T \|^n \sum_{k=0}^{m-n} \| T \|^k \\ &= \| T \|^n \frac{1 - \| T \|^{m-n}}{1 - \| T \|} \\ &\leq \frac{\| T \|^n}{1 - \| T \|} \to 0 \quad \text{as } n \to \infty. \end{aligned} \tag{2.37}$$

Hence $S = u - \lim_{n\to\infty} \sum_{k=0}^{n} T^k$ exists and $S \in B(H)$.

Set $n = 0$ in (2.37) and we obtain

$$\| \sum_{k=0}^{\infty} T^k \| \leq \frac{1}{1 - \| T \|}. \tag{2.38}$$

In the identity

$$\begin{aligned}(I - T^n) &= (I - T)(I + T + T^2 + \cdots + T^{n-1}) \\ &= (I + T + T^2 + \cdots + T^{n-1})(I - T)\end{aligned}$$

take the uniform limit on both sides to obtain

$$\begin{aligned}I &= u - \lim_{n\to\infty} (I - T^n) \\ &= (I - T)(\sum_{k=1}^{\infty} T^k) = (\sum_{k=0}^{\infty} T^k)(I - T).\end{aligned}$$

Thus we conclude that $(I - T)$ is invertible and that $(I - T)^{-1}$ coincides with $\sum_{k=0}^{\infty} T^k$ on the range of $(I - T)$. However, the relation

$$I = (I - T)(\sum_{k=0}^{\infty} T^k) \tag{2.39}$$

implies that $(I - T) : H \to H$ onto. This follows because if y is any element in H then (2.39) implies that there is an $x \in H$ such that

$$y = (I - T)x$$

where

$$x = (\sum_{k=0}^{\infty} T^k)y.$$

Clearly $x \in H$ since T^k: $H \to H$ for all k.

Therefore we have shown that $(I - T)^{-1}$ is defined everywhere on H and that

$$(I - T)^{-1} = \sum_{k=0}^{\infty} T^k. \tag{2.40}$$ ■

An important class of bounded linear operators are the so-called finite dimensional operators.

Definition 2.70

An operator $T \in B(H)$ is said to be **finite dimensional** or an operator of **finite rank** if the linear manifold $T(H)$ is finite dimensional.

An operator $T \in B(H)$ of finite rank has a representation of the form

$$Tf = \sum_{k=1}^{N} (f, y_k)x_k$$

where $x_k, y_k \in H$, $k = 1, 2, \cdots N$ and $\dim T(H) = N$. We notice that $R(T)$, the range of T, is finite dimensional and that $\dim R(T) = N$.

A significant rôle will be played by so-called **compact operators** which have similar properties to those of operators of finite rank.

Definition 2.71

An operator $L \in B(H)$ is **compact** if it is the limit of a uniformly convergent sequence $\{L_k\}_{k=1}^{\infty}$ of operators of finite rank. That is $\| L - L_k \| \to 0$ as $k \to \infty$. An equivalent definition is

Definition 2.71(a)

An operator $L \in B(H)$ is **compact** if for every bounded sequence $\{x_k\}_{k=1}^{\infty} \subset H$ the sequence $\{Lx_k\}_{k=1}^{\infty}$ has a strongly convergent subsequence.

Example 2.72

(i) If L_1 is a compact operator and L_2 a bounded operator on the same Hilbert space H then $L_1 L_2$ and $L_2 L_1$ are also compact operators.

(ii) Let L_k, λ_k, $k = 1, 2, \cdots, N$ be, respectively, compact linear operators and complex valued coefficients, then

$$L = \sum_{k=1}^{N} \lambda_k L_k$$

is a compact operator.

(iii) Any operator of finite rank is compact.

The existence of an inverse operator is of paramount importance in the problem of solving operator equations. Unfortunately not all operators have an inverse. However associated with a given operator on a Hilbert space H is an operator that has some of the flavour of an inverse, namely, an adjoint operator.

Theorem 2.73

Let H be a Hilbert space with structure $(\cdot,\cdot)$ and $T \in B(H)$. Then there exists a unique bounded, linear operator T^* on H, called **adjoint** of T defined by

$$(Tx, y) = (x, T^*y) \text{ for all } x, y \in H$$

such that

$$\| T^* \| = \| T \| .$$

Elementary properties of T^* are contained in

Theorem 2.74

Let H be a Hilbert space and $T, S \in B(H)$. Then

(i) $(T^*)^* =: T^{**} = T$.

(ii) $(\lambda T)^* = \bar{\lambda} T^*, \quad \lambda \in \mathbb{C}$.

(iii) $(T + S)^* = T^* + S^*$.

(iv) $(TS)^* = S^* T^*$.

(v) If T is invertible then so is T^* and

$$(T^*)^{-1} = (T^{-1})^*.$$

(vi) $\| T^* T \| = \| T \|^2$.

Definition 2.75

Let H be a Hilbert space and $T \in B(H)$.

(i) If $T = T^*$ then T is **self-adjoint** or **Hermitian**.

(ii) If $T^* = T^{-1}$ then T is **unitary**.

(iii) If $TT^* = T^*T$ then T is **normal.**

(iv) If T is unitary and $R, S \in B(H)$ are such that

$$R = TST^{-1}$$

then R and S are **unitarily equivalent** with respect to T.

We notice from Theorem 2.73 and Definitions 2.53, 2.67 that

(a) $T \in B(H)$ is self-adjoint if and only if

$$(Tx, y) = (x, Ty) \quad \text{for all } x, y \in H.$$

(b) $T \in B(H)$ is unitary if and only if T is invertible and

$$(Tx, y) = (x, T^{-1}y) \quad \text{for all } x, y \in H.$$

Theorem 2.76

Let H be a Hilbert space with structure $(\cdot,\cdot)$. If $T \in B(H)$ is self-adjoint then

$$\| T \| = \sup_{\|x\|=1} | (Tx, x) | = \max\{| m |, | M |\}$$

where

$$M := \sup\{(Tx, x) \ : \| x \| = 1\}$$
$$m := \inf\{(Tx, x) \ : \| x \| = 1\}$$

are respectively the upper and lower bounds of T.

Definition 2.77

Let H be a Hilbert space with structure $(\cdot,\cdot)$. $T \in B(H)$ and self-adjoint is said to be non-negative if and only if $(Tx, x) \geq 0$ and we write $T \geq 0$. When the inequality is strict we say that T is **positive.**

We remark that if T is self-adjoint then (Tx, x) is real.

It now follows from Theorem 2.76 and Definitions 2.77 that

(a) T is a non negative if $m \geq 0$.

(b) T_1, T_2 are such that $T_1 \leq T_2$ if and only if $T_2 - T_1 \geq 0$.

One of the simplest self–adjoint operators is the projection operator, which is defined in the following way. Let H be a Hilbert space with structure $(\cdot,\cdot)$ and let $M \subset H$ be a closed subspace (that is a closed linear manifold). The Projection Theorem (Theorem 2.54) states that every $x \in H = M \oplus M^{\perp}$ can be expressed uniquely in the form

$$x = y + z, \quad y \in M, z \in M^{\perp}. \tag{2.41}$$

The uniqueness of this representation defines a linear operator

$$\begin{aligned} &P : H \to H \\ &Px = y \in M \end{aligned} \tag{2.42}$$

Such an operator is called a **projection** of H onto M.

We notice that (2.41) can also be written

$$x = y + z = Px + (I - P)x$$

which indicates that

$$(I - P) : H \to H$$

is a projection of H onto $M^{\perp}$.

The main properties of projection operators are contained in the following Theorems.

Theorem 2.78

A bounded linear operator $P : H \to H$ is a projection on the Hilbert space H if and only if

$$P^2 = P = P^*.$$

Theorem 2.79

Let H be a Hilbert space with structure $(\cdot,\cdot)$ and let $M \subset H$ be a closed subspace. Then the projection $P : H \to M$ has the properties

(i) $(Px, x) = \| Px \|^2$.

(ii) $P \geq 0$.

(iii) $\| P \| \leq 1, \quad \| P \| = 1$ if $R(P) \neq \{0\}$.

(iv) $(I - P) =: P^{\perp} : H \to M^{\perp}$

is a projection.

Situations when two or more closed subspaces are involved will be of considerable importance in later chapters. In this connection we have the following results available.

Theorem 2.80

Let H be a Hilbert space with structure $(\cdot,\cdot)$ and let $M_k \subset H$, $k = 1, 2$, be closed subspaces. The projections

$$P(M_k) : H \to M_k, \quad k = 1, 2$$

have the following properties.

(i) The operator $P := P(M_1)P(M_2)$, is a projection on H if and only if

$$P(M_1)P(M_2) = P(M_2)P(M_1)$$

in which case

$$P : H \to R(P) = M_1 \cap M_2.$$

(ii) M_1, M_2 are orthogonal if and only if

$$P(M_1)P(M_2) = 0$$

in which case $P(M_1)$ and $P(M_2)$ are said to be mutually **orthogonal projections.**

(iii) The operator $P := P(M_1) + P(M_2)$ is a projection on H if and only if M_1, M_2 are orthogonal, in which case

$$P := H \to R(P) = M_1 \oplus M_2.$$

(iv) The projections $P(M_1)$, $P(M_2)$ are **partially ordered** in the sense that the following statements are equivalent

(a) $P(M_2)P(M_1) = P(M_1)P(M_2) = P(M_1)$

(b) $M_1 \subseteq M_2$

(c) $N(P(M_2)) \subseteq N(P(M_1))$ where $N(P(M_k))$ denotes the null space of $P(M_k), k = 1, 2$.

(d) $\| P(M_1)x \| \leq \| P(M_2)x \|$ for all $x \in H$

(e) $P(M_1) \leq P(M_2)$.

(v) The operator $P := P(M_2) - P(M_1)$ is a projection on H if and only if $M_1 \subset M_2$ in which case

$$P : H \to R(P) = M_2 \cap M_1^{\perp}.$$

(vi) A series of mutually orthogonal projections $P(M_k), k = 1, 2, \cdots$ on H denoted by $\sum_k P(M_k)$, is strongly convergent to the projection $P(M)$ where $M = \sum_k \oplus M_k$.

(vii) A linear combination of projections $P(M_k)$, $k = 1, 2, \cdots, N$ on H with real valued coefficients λ_k, $k = 1, 2, \cdots, N$, denoted by

$$\sum_{k=1}^{N} \lambda_k P(M_k)$$

is self-adjoint on H.

(viii) Let $\{P(M_k)\}$ denote a monotonically increasing sequence of projections $P(M_k)$, $k = 1, 2, \cdots$ on H. Then

(a) $\{P(M_k)\}$ is strongly convergent to a projection P, that is $P(M_k)x \to Px$ as $k \to \infty$ for all $x \in H$.

(b) $P : H \to R(P)$

$$R(P) = \overline{\bigcup_{k=1}^{\infty} R(P(M_k))}.$$

(c) $N(P) = \bigcap_{k=1}^{\infty} N(P(M_k))$.

Finally in this section dealing with bounded operators on Hilbert spaces we introduce the following notion.

Definition 2.81

Let H be a Hilbert space with structure $(\cdot,\cdot)$. An operator $T \in B(H)$ is called a **Hilbert-Schmidt operator** if

$$\| T \|_{HS}^2 := \sum_{k=1}^{\infty} \| Te_k \|^2 < \infty \tag{2.43}$$

where $\{e_k\}_{k=1}^{\infty}$ is an orthonormal basis for H.

It can be shown that the definition (2.43) is independent of the choice of basis.

Theorem 2.82

(i) Every Hilbert–Schmidt operator is compact.

(ii) $\| T \|_{HS} = \| T^* \|_{HS}$

(iii) $\| T \| \leq \| T \|_{HS}$.

2.6 Unbounded linear operators on Hilbert spaces

Not all linear operators are bounded; for instance the differentiation operator on the space of continuous functions is not bounded, simply consider its action on x^n. However, practically all the operators we encounter in applications are the so-called closed operators and as such they retain much of the flavour of continuity displayed by bounded operators.

Definition 2.83

Let X_1, X_2 be normed linear spaces and $T : X_1 \to X_2$ a linear operator with domain $D(T) \subset X_1$.

If

(i) $x_n \in D(T) \quad \forall n$.

(ii) $x_n \to x \quad$ in X_1.

(iii) $Tx_n \to y \quad$ in X_2.

taken together imply $x \in D(T)$ and $Tx = y$ then T is said to be a **closed operator**.

We would emphasise that in Definition 2.83 we require the **simultaneous** convergence of $\{x_n\}_{n=1}^{\infty}$ and $\{Tx_n\}_{n=1}^{\infty}$.

An alternative definition of closed operator can be given in terms of the **graph** of an operator.

Definition 2.84

Let X_1, X_2 be normed linear spaces with norms $\| \cdot \|_1$, $\| \cdot \|_2$ respectively. A linear operator

$$T : X_1 \supseteq D(T) \to X_2$$

is a closed linear operator if and only if its graph, $G(T)$,

$$G(T) := \{(x_1, x_2) \in X_1 \times X_2 : x_1 \in D(T), x_2 = Tx_1\}$$

is a closed subset of $X_1 \times X_2$.

We would remark that $X_1 \times X_2$ is a normed linear space where the vector space algebraic operations are defined in the usual component–wise manner and the norm, the graph norm, is defined by

$$\| (x_1, x_2) \| = \| x_1 \|_1 + \| x_2 \|_2 . \tag{2.44}$$

These two definitions are equivalent.

In order to establish whether or not a closed operator is in fact a bounded operator we need some if not all of the following results

Theorem 2.85 (Uniform Boundedness Theorem)

Let

(i) I be an index set.

(ii) X, Y be Banach spaces.

(iii) $\{T_\alpha\}_{\alpha \in I} \subset B(X, Y)$ be such that

$$\sup_{\alpha \in I} (\| T_\alpha x \|_Y) < \infty \quad \text{for all } x \in X.$$

Then $\sup_{\alpha \in I} \| T_\alpha \| < \infty$. That is if $\{T_\alpha x\}$ is a bounded set in Y for all $x \in X$ then $\{T_\alpha\}$ is bounded in $B(X, Y)$.

Theorem 2.86a (Bounded Inverse Theorem)

Let

(i) X, Y be Banach spaces

(ii) $T \in B(X, Y)$ be $(1-1)$ and onto.

Then T^{-1} is bounded.

Theorem 2.86b (Closed Graph Theorem)

Let

(i) X, Y be Banach spaces.

(ii) $T : X \to Y$ be a closed linear operator with $D(T) \subset X$.

If $D(T)$ is closed in X then T is bounded.

Example 2.87

We shall show that differentiation can be realised as a closed operator on the space of continuous functions.

Let $X = Y = C[a, b]$, $-\infty < a < b < \infty$, endowed with the usual supremum norm denoted by $\|\cdot\|_\infty$. Define

$$\begin{aligned} &T : X \supseteq D(T) \to Y \\ &(Tg)(x) = \frac{dg}{dx}(x) \equiv g'(x), \quad g \in D(T), x \in [a, b] \\ &D(T) = \{g \in X : g' \in X, g(a) = 0\}. \end{aligned}$$

Let $\{g_n\} \subset D(T)$ be a sequence with the properties that as $n \to \infty$ we have $g_n \to g$ and $Tg_n \to h$ with respect to $\|\cdot\|_\infty$. To establish that T is closed we must show that $g \in D(T)$ and that $Tg = h$. To this end consider

$$f(x) := \int_a^x h(t)dt, \quad x \in [a, b].$$

The convergence $Tg_n \to h$ with respect to $\|\cdot\|_\infty$ implies that the convergence is uniform with respect to x. Since the $Tg_n \in C[a, b]$ it then follows that $h \in C[a, b]$. Consequently by the Fundamental Theorem of the Calculus we deduce that f is continuous and differentiable with $f'(x) = h(x)$ for all $x \in [a, b]$. Furthermore the properties of the

Riemann integral indicate that $f(a) = 0$. Collecting these results we can conclude that $f \in D(T)$.

Since for $g_n \in D(T)$ the Fundamental Theorem of the Calculus implies

$$g_n(x) = \int_a^x g_n'(t)dt,$$

we have

$$\begin{aligned} | g_n(x) - f(x) | &= \left| \int_a^x \{g_n'(t) - h(t)\}dt \right| \\ &\leq \| g_n' - h \|_\infty (b - a). \end{aligned}$$

Now, the right hand side tends to zero by virtue of the convergence $Tg_n \to h$ and it follows that $g_n \to f \in D(T)$. However, since by assumption $g_n \to g$ it then follows that $g = f \in D(T)$. Furthermore, we will also have that $g' = f' = h$. Hence T is defined as a closed operator.

If an operator T is not closed then it is sometimes possible to associate with it an operator which is closed. This parallels the process of associating with a set M in a metric space a closed set $\bar{M}$ called the closure of M.

Definition 2.88

A linear operator T is closable if whenever

(i) $f_n \in D(T)$

(ii) $f_n \to 0$ as $n \to \infty$

(iii) Tf_n tends to a limit as $n \to \infty$

then $Tf_n \to 0 \quad$ as $n \to \infty$.

If T is a closable operator defined on a normed linear space X then we may define an extension of T, denoted by $\bar{T}$ and called the closure of T, in the following manner.

(i) Define

$$D(\bar{T}) = \{f \in X : \exists \{f_n\} \subset D(T) \text{ with } f_n \to f \text{ and } \{Tf_n\} \text{ a Cauchy sequence.}\}$$

Equivalently, we can define $D(\overline{T})$ to be the closure of $D(T)$ with respect to the Graph norm $\| f \|_G^2 := \| f \|_X^2 + \| Tf \|_X^2$.

(ii) For $f \in D(\bar{T})$ set

$$\bar{T}f = \lim_{n\to\infty} Tf_n$$

where f_n is defined as in (i).

That this provides a well defined operator can be seen as follows. If $\{f_n\}$ and $\{g_n\}$ are two sequences each convergent to f then

$$\| f_n - g_n \| \to 0 \quad \text{as } n \to \infty.$$

If $\{Tf_n\}$ and $\{Tg_n\}$ are both Cauchy sequences then

$$\| Tf_n - Tg_n \| \to 0 \quad \text{as } n \to \infty$$

because T is closable by hypothesis. Hence

$$\lim_{n\to\infty} Tf_n = \lim_{n\to\infty} Tg_n$$

which implies that $\bar{T}$ is well defined. We remark that T is closed if $T = \bar{T}$.Connections between closed, bounded and inverse operators are given by

Theorem 2.89

Let X be a Banach space. An operator $T \in B(X)$ is closed iff $D(T)$ is closed.

Theorem 2.90

Let

(i) X_1, X_2 be normed linear spaces

(ii) $T : X_1 \supseteq D(T) \to X_2$, linear, $(1-1)$ operator.

Then T is closed iff

$$T^{-1} : X_2 \supseteq R(T) \to X_1$$

is closed.

The following Theorem draws together a number of results which will be used frequently in later sections.

Theorem 2.91

Let

(i) X_1, X_2 be normed linear spaces,

(ii) $T : X_1 \supseteq D(T) \to X_2$ a linear operator

(a) If there exists a constant $m \geq 0$ such that

$$\| Tx \|_2 \geq m \| x \|_1, \quad x \in D(T)$$

then T is $(1-1)$. Furthermore T is closed iff $R(T)$ is closed in X_2.

(b) Let T be $(1-1)$ and closed. Then $T^{-1} \in B(X_2, X_1)$ iff $R(T)$ is dense in X_2 and there exists $m > 0$ such that

$$\| Tx \|_2 \geq m \| x \|_1, \quad x \in D(T).$$

(c) If T is closed then

$$N(T) := \{x \in D(T) : Tx = 0\}$$

is a closed subset of X_1.

Proof

(a) $Tx = Ty$ for some $x, y \in D(T)$ implies

$$\| x - y \|_1 \leq m^{-1} \| T(x-y) \|_2 = m^{-1} \| Tx - Ty \|_2 = 0.$$

Hence $x = y$ which implies T is $(1-1)$. Therefore, for any $y \in D(T^{-1})$ where

$$D(T^{-1}) = \{y \in R(T) : \exists x \in D(T) \text{ s.t. } y = Tx\}$$

we have

$$\| x \|_1 \leq m^{-1} \| Tx \|_2$$

and by definition of $D(T^{-1})$

$$\| T^{-1}y \| \leq m^{-1} \| y \|_2$$

which implies that T^{-1} is bounded on $R(T)$. Hence, using Theorems 2.89 and 2.90 we obtain

$R(T)$ is closed in $X_2 \Leftrightarrow T^{-1}$ closed $\Leftrightarrow T$ closed.

(b) ($\Rightarrow$) $T^{-1} \in B(X_2, X_1)$ implies that $R(T) = D(T^{-1}) = X_2$ and that there is a constant $M > 0$ such that

$$\| T^{-1}y \|_1 \leq M \| y \|_2, \quad y \in X_2. \tag{2.45}$$

Since by hypothesis T^{-1} exists then there exists an $x \in X_1$ such that $x = T^{-1}y$. Consequently (2.45) yields

$$\| x \|_1 \leq M \| Tx \|_2, \quad x \in D(T)$$

and the required result follows by setting $m = M^{-1}$.

(b) ($\Leftarrow$) If $R(T)$ is dense in X_2 then since T is $(1-1)$ and closed it follows from (a) that $R(T)$ is closed. Hence

$$R(T) = \overline{R(T)} = X_2$$

the last equality following by hypothesis. Thus T is $(1-1)$ and onto X_2 and hence T^{-1} exists. Since we also have, by hypothesis that there exists an $m > 0$ such that

$$\| Tx \|_2 \geq m \| x \|_1, \quad x \in D(T)$$

then setting $y = Tx$ indicates that $T^{-1} \in B(X_2, X_1)$ as required.

(c) Let x be a limit point of $N(T)$. Then there exists a sequence $\{x_n\} \subset N(T)$ such that $x_n \to x$ as $n \to \infty$. Furthermore, since $Tx_n = 0$ for all n we can write $Tx_n \to 0$ as $n \to \infty$. Consequently, since T is closed we have $x \in D(T)$ and $Tx = 0$ which implies that $x \in N(T)$. Therefore we have that $\overline{N(T)} \subset N(T)$. However, since trivially $N(T) \subseteq \overline{N(T)}$, it follows that $N(T) = \overline{N(T)}$ and that $N(T)$ is closed. ■

2.7 Adjoints of unbounded operators

A particularly useful feature of a bounded linear operator T on a Hilbert space H is that it has associated with it a bounded linear operator T^*, the adjoint of T, which is defined by the equation

$$(Tf, g) = (f, T^*g) \quad \text{for all } f, g \in H \tag{2.46}$$

where $(\cdot, \cdot)$ is the inner product on H. (See Theorem 2.73). The proof of the existence of such an adjoint operator, which uses the Riesz Representation Theorem, breaks down when either T is unbounded or not defined on all of H. Nevertheless, even when the proof breaks down it may happen that for some element $g \in H$ there is an element $g^* \in H$ such that

$$(Tf, g) = (f, g^*) \quad \text{for all } f \in D(T). \tag{2.47}$$

If for some fixed $g \in H$ there is only one $g^* \in H$ such that (2.47) holds then we can write $g^* = T^*g$ and consider the operator T^* as being well defined for at least this particular $g \in H$. However, it remains to determine the conditions under which (2.47) yields a unique solution $g^* \in H$. Results in this direction are given by

Theorem 2.92

Let

(i) T be a linear operator on a Hilbert space H

(ii) there exist elements $g, g^* \in H$ such that

$$(Tf, g) = (f, g^*) \quad \text{for all } f \in D(T)$$

then g^* is uniquely determined by g and (2.47) iff $D(T)$ is dense in H.

Theorem 2.93

Let

(i) T be a linear operator on a Hilbert space H with $\overline{D(T)} = H$

(ii) $D(T^*) = \{g \in H : \text{there exists } g^* \in H \text{ satisfying (2.47)}\}$.

then $D(T^*)$ is a subspace of H and the operator T^*, with domain $D(T^*)$, defined by

$$T^*g = g^* \quad \text{for all } g \in D(T^*)$$

is a linear operator.

These two results lead naturally to the following definition.

Let T be a linear operator in H with $\overline{D(T)} = H$. The linear operator T^* with domain $D(T^*)$ as defined in Theorem 2.93 is called the **adjoint** of T.

When T is a bounded operator then this definition of an adjoint operator coincides with that given in Theorem 2.73. However, in the general case more attention must be given to the role and importance of domains of operators. The results given below illustrate this fact.

Theorem 2.94 (properties of adjoints)

Let T, S be linear operators on a Hilbert space H.

(i) If $T \subset S$ and $\overline{D(T)} = H$ (which also implies $\overline{D(S)} = H$) then $T^* \supset S^*$.

(ii) If $\overline{D(T)} = \overline{D(T^*)} = H$ when $T \subset T^{**}$.

(iii) If T is $(1-1)$ and such that $\overline{D(T)} = \overline{D(T^{-1})} = H$ then T^* is $(1-1)$ and

$$(T^*)^{-1} = (T^{-1})^*.$$

The following types of operator are important in applications.

Definition 2.95

Let T be a linear operator in a Hilbert space H such that $\overline{D(T)} = H$.

(i) If $T^* = T$ then T is **self-adjoint.**

(ii) If $T^* \supset T$ then T is **symmetric.**

Some properties of these operators are collected together in the following

Theorem 2.96

Let T be a linear operator in a Hilbert space H.

(i) If T is $(1-1)$ and self-adjoint then $\overline{D(T^{-1})} = H$ and T^{-1} is self-adjoint.

(ii) If T is defined everywhere on H then T^* is bounded.

(iii) If T is self-adjoint and defined everywhere on H then T is bounded.

(iv) If $\overline{D(T)} = H$ then T^* is closed.

(v) Every self–adjoint operator is closed.

(vi) If T is closable then $(\bar{T})^* = T^*$ (and $\overline{T} = T^{**}$).

(vii) If T is symmetric then it is closable.

2.8 A basic criterion for self-adjointness

Self-adjoint operators have a particularly important role to play in subsequent sections and here we shall indicate a means of deciding whether or not a given operator has this property. First we introduce

Definition 2.97

Let T be a linear operator in a Hilbert space H with $\overline{D(T)} = H$. The operator T is said to be **essentially self-adjoint** if $\bar{T} = T^*$.

We notice that if T is essentially self-adjoint then necessarily it must be closable, since T^* is a closed operator, and from Theorem 2.95 (vi) we conclude that $T^* = \bar{T}^*$. Therefore T is essentially self-adjoint if and only if

$$\bar{T} = \bar{T}^*.$$

That is, if and only if $\bar{T}$ is self-adjoint.

We would also emphasise that a symmetric operator need not be self-adjoint. However, it might have a self-adjoint extension and then again it might not. We shall be particularly interested in the case when a symmetric operator has exactly one self-adjoint extension, that is, when it is essentially self-adjoint. In this connection we shall first establish the following.

Theorem 2.98

Let T be a linear, symmetric operator in a Hilbert space H, Then

(i) T is closable.

(ii) $R(\bar{T} \pm iI) = \overline{R(T \pm iI)}$.

Proof: (i) T symmetric $\Rightarrow$ T has a closed extension (namely T^*).

Assume $\{f_n\} \subset D(T)$ is such that

$$s - \lim_{n\to\infty} f_n = 0 \quad \text{and} \quad s - \lim_{n\to\infty} Tf_n = h.$$

Since $T \subseteq T^*$ then

$$Tf_n = T^* f_n \quad \forall f_n \in D(T).$$

Hence

$$s - \lim_{n\to\infty} T^* f_n = s - \lim_{n\to\infty} T f_n = h.$$

Since T^* is a closed operator (Theorem 2.96) then

$$h = s - \lim_{n\to\infty} T^* f_n = T^*(s - \lim_{n\to\infty} f_n) = 0$$

and it follows that T is closable.

(ii) Let $f \in D(\bar{T})$. This implies that there exists $\{f_n\} \subset D(T)$ such that (see Definition 2.88)

$$f = s - \lim_{n\to\infty} f_n \quad \text{and} \quad \bar{T} f = s - \lim_{n\to\infty} T f_n.$$

Thus

$$s - \lim_{n\to\infty} (T \pm iI) f_n = (\bar{T} \pm iI) f$$

which implies that

$$R(\bar{T} \pm iI) = (\bar{T} \pm iI) D(\bar{T}) \subseteq \overline{R(T \pm iI)}.$$

To obtain the reverse inclusion we notice that T symmetric and $f \subset D(T)$ yields

$$\begin{aligned} \| (T \pm iI) f \|^2 &= ((T \pm iI) f, (T \pm iI) f) \\ &= \| Tf \|^2 + \| f \|^2 . \end{aligned} \tag{2.48}$$

Now suppose $g \in \overline{(T \pm iI) D(T)} = \overline{R(T \pm iI)}$ then there exists $\{f_n\} \subset D(T)$ such that

$$\lim_{n\to\infty} (T \pm iI) f_n = g. \tag{2.49}$$

Therefore, using (2.48) we obtain

$$\begin{aligned} \| f_n - f_m \|^2 &= \| (T \pm iI)(f_n - f_m) \|^2 - \| T(f_n - f_m) \|^2 \\ &\leq \| (T \pm iI)(f_n - f_m) \|^2 \end{aligned} \tag{2.50}$$

and the right-hand side tends to zero as $n, m \to \infty$ as a consequence of (2.49).

Similarly

$$\begin{aligned} \| Tf_n - Tf_m \|^2 &= \| (T \pm iI)(f_n - f_m) \|^2 - \| f_n - f_m \|^2 \\ &\leq \| (T \pm iI)(f_n - f_m) \|^2 \end{aligned} \tag{2.51}$$

and again the right-hand side tends to zero as $n, m \to \infty$ by (2.49).

Consequently, $\{f_n\}$ and $\{Tf_n\}$ are Cauchy sequences so, by definition of closure (Definition 2.88)

$$h := s-\lim_{n\to\infty} f_n \in D\overline{T}$$

and $\overline{T}h = s-\lim_{n\to\infty} Tf_n$. Therefore

$$\begin{aligned} g &= s-\lim_{n\to\infty}(T \pm iI)f_n = s-\lim_{n\to\infty} Tf_n + s-\lim_{n\to\infty} if_n \\ &= \overline{T}h \pm ih = (\overline{T} \pm iI)h. \end{aligned}$$

That is, $g \in R(\overline{T} \pm iI)$. Combining this result with (2.49) we obtain

$$\overline{R(T \pm iI)} \subseteq R(\bar{T} \pm iI)$$

the required reverse inclusion. The Theorem follows. ■

This Theorem will now be used to prove the main result of this section.

Theorem 2.99

A symmetric operator $T : H \to H$ is self-adjoint if and only if

$$\{f : f = (T + iI)g, g \in D(T)\} = H = \{f : f = (T - iI)g, g \in D(T)\}$$

that is, if and only if

$$R(T + iI) = H = R(T - iI).$$

Proof: Assume

(i) $(T \pm iI)D(T) = R(T \pm i) = H$.

(ii) $g \in D(T^*)$.

Then, since $(T - iI)D(T) = H$ there will exist an $h \in D(T)$ which will allow us to write

$$(T^* - iI)g = (T - iI)h. \tag{2.52}$$

Furthermore, since $T \subseteq T^*$ then $D(T) \subseteq D(T)^*$ and

$$(T - iI)h = (T^* - iI)h, \quad h \in D(T). \tag{2.53}$$

Combining (2.52) and (2.53) we obtain

$$(T^* - iI)g = (T^* - iI)h$$

that is

$$(T^* - iI)(g - h) = 0.$$

Now, for any $f \in D(T)$ we have

$$0 = ((T^* - iI)(g - h), f) = (g - h, (T + iI)f) \tag{2.54}$$

and since $(T + iI)D(T) = H$ then (2.54) holds for all elements in H which implies

$$g - h = 0.$$

Therefore, $g = h \in D(T)$ which implies that $D(T^*) \subseteq D(T)$. Since we already know that $D(T) \subseteq D(T^*)$ it follows that $D(T) = D(T^*)$ and we can conclude that

$$T = T^*.$$

Conversely, assume that $T = T^*$.

Since T is self-adjoint it is symmetric and by Theorem 2.98 has the property that

$$\overline{R(T \pm iI)} = R(\bar{T} \pm iI).$$

Since a self-adjoint is closed, $T = \bar{T}$ and we conclude that

$$R(T \pm iI) = (T \pm iI)D(T)$$

are closed subspaces of H. If it is the case that

$$R(T \pm iI) \neq H$$

then there exists a non-trivial element $g \in R(T \pm iI)^{\perp}$ and it follows that

$$\begin{aligned} 0 &= ((T \pm iI)f, g) \quad \forall f \in D(T \pm iI) \\ &= (f, z) \end{aligned} \tag{2.55}$$

where

$$z := (T^* \mp iI)g = (T \mp iI)g.$$

From (2.55) we conclude

$$g \in D((T \pm iI)^*) = D(T \mp iI) = D(T).$$

Also from (2.55) we infer

$$z = (T^* \mp iI)g = 0.$$

Therefore

$$\begin{aligned}(g, g) &= (g, \mp iT^* g) \\ &= (\pm iTg, g) = (\pm iT^* g, g) \\ &= (-g, g)\end{aligned}$$

which implies $g = 0$ and hence

$$R(T \pm iI)^{\perp} = \{0\}.$$

Thus

$$R(T \pm iI) = H. \qquad \blacksquare$$

Chapter 3
Examples of scattering theory strategies

3.1 Introduction

Scattering theory compares the behaviour in the distant past and the distant future of a system evolving in time. The comparison is made relative to some free or unperturbed problem. The given problem, regarded as the perturbed problem, is then measured against the free problem. Specifically, we investigate whether or not solutions of the perturbed problem behave like solutions of the unperturbed problem in either the distant past or distant future or indeed both; that is we enquire into the **asymptotic equality** of the solutions.

In order to be able to make these comparisons we must, of course, be confident that the various solutions actually exist. To illustrate how this aspect might be approached consider systems governed by initial boundary value problems of the form

$$\begin{aligned}
&\left(\frac{\partial^2}{\partial t^2} - L\right) u(x,t) = f(x,t), \quad (x,t) \in \Omega \times \mathbf{R} \\
&u(x,0) = u_0(x), \quad u_t(x,0) = u_1(x), \quad x \in \Omega \qquad\qquad \mathbf{P1} \\
&u(x,t) \in (bc), \quad (x,t) \in \partial\Omega \times \mathbf{R}
\end{aligned}$$

where L is a linear second order differential expression defined on $\Omega \subset \mathbf{R}^n$, an open, connected region with smooth boundary $\partial\Omega$. The function f represents the energy sources in the systems. The notation (bc) denotes that the solution, u, is required to satisfy boundary conditions on $\partial\Omega$. Together with f and (bc) the initial data u_0 and u_1 are also known.

The first thing we have to do when analysing a problem is to declare the type of solution we are actually looking for; that is, we must introduce a **solution concept.** For the purposes of this particular illustration we shall investigate whether or not there are solutions of **P1** which exist in the Hilbert space $L_2(\Omega)$ endowed with the usual structure. Consequently, if we introduce the linear operator A defined by

$$\begin{aligned}
&A : L_2(\Omega) \to L_2(\Omega) \\
&Au = -Lu, \quad u \in D(A)
\end{aligned}$$

$$D(A) = \{u \in L_2(\Omega) : Lu \in L_2(\Omega), \quad u \in (bc)\}$$

then **P1** has in $L_2(\Omega)$, the operator realisation

$$\left(\frac{\partial^2}{\partial t^2} + A\right) u(x,t) = f(x,t), \quad (x,t) \in \Omega \times \mathbf{R}$$
$$u(x,0) = u_0(x), \quad u_t(x,0) = u_1(x), \quad x \in \Omega \qquad \mathbf{P2}$$

where we notice that the boundary conditions (bc) have been absorbed into the definition of A.

If now we regard u in **P2** as an n-times continuously differentiable mapping from $\mathbf{R}$ into $L_2(\Omega)$, a situation we denote by $u \in C^n(\mathbf{R}, L_2(\Omega))$, then we can write

$$u : \mathbf{R} \to L_2(\Omega) \qquad u : t \to u(\cdot, t) \in L_2(\Omega).$$

With this understanding **P2** reduces to an ordinary differential equation problem of the form

$$\left(\frac{d^2}{dt^2} + A\right) u = f, \quad u(0) = u_0, \quad u_t(0) = u_1. \qquad \mathbf{P3}$$

In **P3** the quantity A is now regarded as a constant and we obtain, as a consequence, a solution of the form

$$u(\cdot,t) = (\cos tA^{\frac{1}{2}})u_0 + A^{-\frac{1}{2}}(\sin tA^{\frac{1}{2}})u_1 + \\ + \int_0^t A^{-\frac{1}{2}}(\sin(t-s)A^{\frac{1}{2}})f(s)ds \qquad (I).$$

In (I) the first two terms comprise the Complementary Function for **P3** whilst the third term is a Particular Integral obtained in terms of the Greens Function for **P3**. The solution, $u(x,t)$, of **P1** can now be recovered from (I) provided $A^{\pm\frac{1}{2}}, \cos tA^{\frac{1}{2}}$ and $\sin tA^{\frac{1}{2}}$ can be meaningfully defined. We shall see in Chapter 4 that this can be done by using the Spectral Theorem and the properties of spectral families.

Alternatively, **P3** can be reduced to a first order system by setting $u_t = v$. We then obtain

$$\begin{bmatrix} u \\ v \end{bmatrix}_t + \begin{bmatrix} 0 & -1 \\ A & 0 \end{bmatrix} \begin{bmatrix} u \\ v \end{bmatrix} = \begin{bmatrix} 0 \\ f \end{bmatrix}, \quad \begin{bmatrix} u \\ v \end{bmatrix}(0) = \begin{bmatrix} u_0 \\ u_1 \end{bmatrix}.$$

This can be written more conveniently, with an obvious notation, as

$$\psi_t + G\psi = F, \quad \psi(0) = \psi_0 \qquad \textbf{P4}$$

where it is understood that

$$G : L_2(\Omega) \times L_2(\Omega) \to L_2(\Omega \times L_2(\Omega).$$

Using the integrating factor technique to solve the ordinary differential equation in **P4** we obtain

$$\psi(t) = U(t - t_0)\psi(t_0) + \int_{t_0}^{t} U(t - s)F(s)ds \qquad (II)$$

where $U(t) := e^{-tG}$ and t_0 is arbitrary.

The relation (II) represents a solution of **P4** which has initial data given at $t = t_0$. In our particular case $t_0 = 0$ and we obtain

$$\psi(t) = U(t)\psi_0 + \int_0^t U(t - s)F(s)ds. \qquad (III)$$

The steps leading to (II) and (III) can of course be justified; the main tool we shall use for doing this will be semigroup theory which will be introduced in Chapter 5. We shall see that the abstract results of semigroup theory can be used to furnish powerful results concerning the existence and uniqueness of solutions to the initial boundary value problems in which we are interested. Furthermore, they will also enable us to discuss perturbation effects in an elegant manner. However, an immediate consequence of the theory will be seen to be that $U(t) = \exp(-tG)$ is well defined provided G is self-adjoint.

We have already mentioned that the representation (I) can be justified and made constructive by first establishing that A is self-adjoint in the chosen underlying function space and then using the Spectral Theorem. To use this Theorem we need to determine the so-called **spectral family** of A and to do this we need information about $\sigma(A)$, the spectrum of A, and about $R_\lambda(A)$ the resolvent of A. The latter can be obtained provided the Green's function associated with A can be determined. An example of this is given towards the end of Chapter 4.

We would remark that these lines of approach are not by any means confined to an investigation of linear problems. On the contrary, for nonlinear problems we shall see that we can still obtain expressions of the form (I) and (II) but now they will be integral equations which must be solved to yield required solutions. However, for the moment we shall confine attention to linear problems.

In developing a scattering theory we have seen that, typically, we investigate solutions to problems of the form

$$u_{tt} + A_1 u = 0, \quad u(0) = u_0, \quad u_t(0) = u_1 \qquad \textbf{P5}$$

$$v_{tt} + A_2 v = 0, \quad v(0) = v_0, \quad v_t(0) = v_1 \qquad \textbf{P6}$$

where $u, v \in C^n(\mathbf{R}, L_2(\Omega))$ and $A_j : L_2(\Omega) \to L_2(\Omega)$, $j = 1,2$, $\Omega \subset \mathbf{R}^n$. From the observations made above we know that we can, at least in principle, settle questions of existence and uniqueness of solutions and of perturbation effects centred on **P5** and **P6**. The development of a scattering theory associated with **P5** and **P6** runs along the following lines.

Given arbitrary initial data (u_0^-, u_1^-) for **P5** which yields a solution $u^-(t)$ show that there exists a solution $v(t)$ of **P6** which behaves asymptotically as $t \to -\infty$ like a solution $u^-(t)$ of **P5** in the sense that

$$\| v(t) - u^-(t) \|_E \to 0 \text{ as } t \to -\infty$$

where $\| \cdot \|_E$ denotes a, so-called , energy norm, which is introduced to confine attention to solutions with finite energy. Furthermore, show that also there exist solutions $u^+(t)$ of **P5** with initial data (u_0^+, u_1^+) such that

$$\| v(t) - u^+(t) \|_E \to 0 \text{ as } t \to +\infty.$$

The mapping

$$S : (u_0^-, u_1^-) \to (u_0^+, u_1^+)$$

is called the **scattering operator** associated with A_1 and A_2. We shall see that the scattering operator can be decomposed in the form

$$S = \Omega_+^{-1} \Omega_-$$

where

$$\Omega_- : (u_0^-, u_1^-) \to (v_0, v_1)$$
$$\Omega_+ : (u_0^+, u_1^+) \to (v_0, v_1)$$

are the **wave operators** associated with A_1 and A_2.

In developing a scattering theory we are principally concerned with the existence and construction of the wave operators $\Omega_\pm$ and with their **asymptotic completeness.** This latter property is focussed on the availability of the data $(u_0^\pm, u_1^\pm)$. A way in which $\Omega_\pm$ can be determined is illustrated in the next sections.

3.2 A free problem

We take as a free problem the initial value problem

$$u_t(x,t) + u_x(x,t) = 0, \quad x \in \mathbf{R}, \quad t \in \mathbf{R} \tag{3.1}$$
$$u(x,0) = f(x), \quad x \in \mathbf{R}. \tag{3.2}$$

We see, by inspection, that this problem has a solution

$$u(x,t) = f(x-t), \quad x \in \mathbf{R}, \quad t \in \mathbf{R} \tag{3.3}$$

which represents a wave travelling in the positive x-direction with unit velocity.

An alternative method for obtaining a solution to (2.1), (2.2) is to use the Fourier transforms F, F^* defined by

$$(Ff)(k) \equiv \hat{f}(k) = \frac{1}{\sqrt{2\pi}} \int_{\mathbf{R}} e^{-ikx} f(x)dx \tag{3.4}$$
$$f(x) = (F^*\hat{f})(x) = \frac{1}{\sqrt{2\pi}} \int_{\mathbf{R}} e^{ikx} \hat{f}(k)dk \tag{3.5}$$

where $F^* = F^{-1}$ is the inverse Fourier transform.

Taking the Fourier transform of the problem (3.1), (3.2) we obtain

$$\hat{u}_t(k,t) + ik\hat{u}(k,t) = 0, \quad k \in \mathbf{R}, \quad t \in \mathbf{R} \tag{3.6}$$
$$\hat{u}(k,0) = \hat{f}(k), \quad k \in \mathbf{R}. \tag{3.7}$$

This ordinary differential equation (3.6) has a solution satisfying (3.7) given by

$$\hat{u}(k,t) = e^{-ikt}\hat{f}(x).$$

The Fourier inversion theorem (3.4), (3.5) now enables us to obtain

$$u(x,t) = \frac{1}{\sqrt{2\pi}} \int_{\mathbf{R}} e^{ikx} \hat{u}(k,t)dk = \frac{1}{\sqrt{2\pi}} \int_{\mathbf{R}} e^{ik(x-t)} \hat{f}(k)dk$$
$$= f(y), \quad y = x - t.$$

Hence we recover the solution (3.3).

Yet another method for obtaining a solution to (3.1), (3.2) and one which is more in the spirit adopted in later sections, is to represent the problem (3.1), (3.2) as a problem in a Hilbert space, for example in $L_2(\mathbf{R})$. To this end we write (3.1) in the form

$$0 = u_t(x,t) + u_x(x,t) = u_t(x,t) + i(-iu_x(x,t)) = u_t(x,t) + iL_0u(x,t) \tag{3.8}$$

where

$$L_0 := -i\frac{\partial}{\partial x}. \tag{3.9}$$

We now introduce an operator A_o defined by

$$\begin{aligned} &A_0 : L_2(\mathbf{R}) \to L_2(\mathbf{R}) =: H \\ &A_0 u = L_0 \quad \forall\, u \in D(A_0) \\ &D(A_0) = \{u \in H : L_0 u \in H\}. \end{aligned} \tag{3.10}$$

Consequently, the initial value problem (3.1), (3.2) can be given the following operator realisation in H. Find $u \in H$ such that

$$u_t + iA_0 u = 0 \quad u \in D(A_0) \tag{3.11}$$

$$u(0) = f \in H \tag{3.12}$$

where u must now be interpreted as an $L_2(\mathbf{R})$-valued function of t. The problem (3.11), (3.12) has a solution which can be written in the form

$$u(x,t) = \exp(-itA_0)f(x) =: U_0(t)f(x) \tag{3.13}$$

which clearly indicates the evolution of the initial data, f, into the solution, u.

However, for (3.13) to be meaningful we require an interpretation of

$$U_0(t) := \exp(-itA_0). \tag{3.14}$$

In the following section we examine certain properties of the operator A_0 and show how they lead to an interpretation of (3.14); this in turn will allow the solution (3.3) to be recovered.

3.3 Properties of the operators A_0 and $U_0(t)$

The operator A_0 is defined in (3.10) where $H \equiv L_2(\mathbf{R})$ has the usual structure

$$(f, g) = \int_{\mathbf{R}} f(x)\overline{g(x)}dx, \quad \| f \|^2 = (f, f). \tag{3.15}$$

(i) A_0 is formally self adjoint:

$$\begin{aligned}(f, A_0 g) &= \int_{-\infty}^{\infty} f(x)\overline{\{-i\frac{\partial g}{\partial x}\}}dx \\ &= i[f(x)\overline{g(x)}]_{-\infty}^{\infty} + \int_{-\infty}^{\infty} \{-i\frac{\partial f}{\partial x}\}\overline{g(x)}dx.\end{aligned}$$

Thus provided

$$\lim_{|x|\to\infty} f(x)\overline{g(x)} = 0 \tag{3.16}$$

then

$$(f, A_0 g) = (A_0 f, g)$$

indicating that A_0 is formally self-adjoint. The operator A_0 is, in fact, self-adjoint since for elements in H the limit (3.16) can be satisfied. This wil be made more precise later.

The operator A_0 assumes a particularly simple form after application of Fourier transforms. With the Fourier transform defined as in (3.4), (3.5) and taking

$$f \in L_2(\mathbf{R}) \cap L_1(\mathbf{R}) \tag{3.17}$$

then it is known from the general theory of Fourier transforms that

(a) $F : \mathbf{R} \to \mathbf{R}$

(b) F is unitary

(c) $(f, g) = \int_{\mathbf{R}} f(x)\overline{g(x)}dx = \int_{\mathbf{R}} \hat{f}(k)\overline{\hat{g}(k)}dk = (\hat{f}, \hat{g})$.

Consequently

$$
\begin{aligned}
F(A_0 f)(x) &= \frac{1}{\sqrt{2\pi}} \int_{\mathbf{R}} e^{-ikx}\{-i\frac{df}{dx}\}dx \\
&= \frac{1}{\sqrt{2\pi}}\{[-if(x)e^{ikx}]_{-\infty}^{\infty} + i\int_{-\infty}^{\infty} \frac{d}{dx}e^{-ikx}f(x)dx\} \\
&= \frac{k}{\sqrt{2\pi}} \int_{-\infty}^{\infty} e^{-ikx}f(x)dx = k\hat{f}(k). \qquad (3.18)
\end{aligned}
$$

The integrated terms vanish because for f as in (3.17) we always have

$$\lim_{|x|\to\infty} f(x) = 0.$$

Thus, from (3.18)

$$F(A_0 f)(k) = k\hat{f}(k) = (kFf)(k)$$

which in turn implies

$$A_0 f(x) = (F^{-1}kF)f(x) = (F^*kF)f(x).$$

Consequently since F is a unitary operator we see that A_0 is unitarily equivalent to the operator of multiplication by k on H. Thus, F diagonalises the operator A_0 (see Chapter 4). We shall also say that F provides a spectral transformation for A_0.

The next two properties also anticipate topics which will be fully discussed in the next Chapter.

(ii) A_0 has no eigenvalues.

The eigenvalue problem

$$A_0 f(x) = -i\frac{df}{dx} = \lambda f(x) \qquad (3.19)$$

has the non-trivial solution

$$f(x) = ae^{i\lambda x}.$$

However

$$\| f \|^2 = \int_{-\infty}^{\infty} | f(x) |^2 \, dx = \int_{-\infty}^{\infty} | ae^{i\lambda x} |^2 \, dx = \int_{-\infty}^{\infty} | a |^2 \, dx$$

which implies that for all values of λ

$$f \notin L_2(\mathbf{R}).$$

Thus A_0 has no eigenvalues. However, it may well have generalised eigenvalues (see Definition 4.59).

(iii) A_0 has a purely continuous spectrum

Applying the Fourier transform to (3.19) we obtain

$$k\hat{f}(k) = \lambda \hat{f}(k). \tag{3.20}$$

Since k can assume all values in $\mathbf{R}$ there will always exist a $\lambda \in (-\infty, \infty)$ such that (3.20) holds.

(iv) A_0 is an unbounded operator.

Let χ_n= characteristic function for the interval $[n, n+1]$ that is

$$\chi_n(k) = \begin{cases} 1, & n \leq k \leq n+1 \\ 0, & \text{otherwise.} \end{cases} \tag{3.21}$$

Then

$$\| \chi_n \|^2 = \int_{-\infty}^{\infty} | \chi_n(k) |^2 \, dk = \int_n^{n+1} dk = 1 \tag{3.22}$$

$$\| k\chi_n \|^2 = \int_{-\infty}^{\infty} | k\chi_n(k) |^2 \, dk = \int_n^{n+1} k^2 dk \geq n^2 \int_n^{n+1} dk = n^2. \tag{3.23}$$

If now we recall the property (c) of Fourier transforms then we have

$$\| \widehat{A_0 f} \|^2 = \| A_0 f \|^2 = \| k\hat{f} \|^2 = \| kf \|^2 . \tag{3.24}$$

Consequently, given any $M < \infty$ there exists $f \in H$ such that $\| f \| = 1$ and $\| \widehat{A_0 f} \| > M$. To see this simply take $f = \chi_m$ and use (3.22) to (3.24).

(v) Group properties.

We have defined

$$U_0(t) = \exp(-itA_0). \tag{3.14}$$

Expanding the right hand side yields

$$\exp(-itA_0) = 1 - itA_0 + \frac{1}{2!}(-itA_0)^2 + \cdots .$$

Consequently

$$\begin{aligned}(FU_0(t)f)(k) &= F([1 - itA_0 + \frac{1}{2!}(-itA_0)^2 + \cdots]f)(k) \\ &= (\hat{f} - itk\hat{f} + \frac{1}{2!}(-itA_0)^2\hat{f} + \cdots)(k) \\ &= \exp(-itk)\hat{f}(k). \qquad (3.25)\end{aligned}$$

Since $\exp(-ikt)\exp(-ik\tau) = \exp(ik[t+\tau])$ it follows from (3.25), by considering $U_0(t)U_0(\tau)$ and $U_0(t+\tau)$, that

$$U_0(t)U_0(\tau) = U_0(t+\tau) \qquad (3.26)$$

a result which implies that $\{U_0(t)\}_{t\in\mathbf{R}}$ is a one-parameter group.

Furthermore

$$U_0(t)U_0(-t) = U_0(-t)U_0(t) = U_0(0) = I$$

which implies that $U(t)$ is invertible with

$$U_0(t)^{-1} = U_0(-t). \qquad (3.27)$$

We also notice that

$$\| U_0(t)f \|^2 = \| (U_0(t)f)^\wedge \|^2 = \int_{-\infty}^{\infty} | e^{-ikt}\hat{f} |^2 \, dk = \| \hat{f} \|^2 = \| f \|^2 .$$

Hence $U_0(t)$ is a unitary operator on H and $\{U_0(t)\}_{t\in\mathbf{R}}$ is a one-parameter unitary group of operators on H.

(vi) Action of $U_0(t)$.

Using the inverse Fourier transform (3.5) we have

$$\begin{aligned}(U_0(t)f)(x) &= \frac{1}{\sqrt{2\pi}}\int_{-\infty}^{\infty} e^{ikx}((U(t)f))^\wedge(k)dk \\ &= \frac{1}{\sqrt{2\pi}}\int_{-\infty}^{\infty} e^{ikx}e^{-ikt}\hat{f}(k)dk \\ &= f(x-t). \qquad (3.28)\end{aligned}$$

Thus $U_0(t)$ induces a translation of argument by an amount t.

Combining (3.13) and (3.28) we see that the solution of the free problem is given by

$$u(x,t) = U_0(t)f(x) = f(x-t). \qquad (3.29)$$

Hence we again recover the solution (3.3).

3.4 A perturbed problem

We now consider the initial value problem

$$v_t(x,t)+v_x(x,t)+q(x)v(x,t)=0, \quad x\in\mathbf{R}, \quad t\in\mathbf{R} \tag{3.30}$$

$$v(x,0)=g(x), \quad x\in\mathbf{R} \tag{3.31}$$

which we consider to be a perturbation of problem (3.1), (3.2).

Here we assume that the potential q satisfies

(i) q is a real valued, bounded function of $x\in\mathbf{R}$

(ii) $q\in L_1(\mathbf{R})$.

Just as for the free problem we can represent the perturbed problem in the Hilbert space $L_2(\mathbf{R})$. To do this we introduce the operator A defined by

$$\begin{aligned} A &: L_2(\mathbf{R})\to L_2(\mathbf{R})\equiv H \\ Av &= \{-i\frac{\partial}{\partial x}+q(x)\}v=(A_0+q(x))v \quad \forall\, v\in D(A) \\ D(A) &= \{v\in D(A_0): qv\in H\}. \end{aligned} \tag{3.32}$$

The initial value problem (3.30), (3.31) has the following realisation in H. Find $v\in H$ such that

$$v_t+iAv=0, \quad v\in D(A) \tag{3.33}$$

$$v(0)=g\in H \tag{3.34}$$

where v must be interpreted as an $L_2(\mathbf{R})$-valued function of t.

Arguing as for the free problem we see that the solution of (3.33), (3.34) can be written in a form which indicates the evolution of the initial data, g, into the solution, v; specifically we have

$$v(x,t)=\exp(-itA)g(x)=:U(t)g(x). \tag{3.35}$$

Again we are faced with the task of interpreting $U(t)$. In this case it is not so easy to work out $U(t)$, it depends crucially on the influence of the potential q appearing in the differential expression

$$L = L_0 + q. \tag{3.36}$$

However we make headway by means of the following results.

(i) A is formally self-adjoint.

Since q is real valued this result follows just as for A_0.

(ii) A and A_0 are unitarily equivalent.

Define

$$\rho(x) := \int_0^x q(y)dy \tag{3.37}$$

$$(Wf)(x) := e^{-i\rho(x)}f(x). \tag{3.38}$$

Since

$$(W^{-1}f)(x) = e^{i\rho(x)}f(x) = (W^*f)(x)$$
$$\| Wf \|^2 = \| e^{-i\rho(x)}f \|^2 = \| f \|^2 .$$

it follows that W is a unitary operator.

Furthermore

$$\begin{aligned}(WA_0W^{-1}f)(x) &= e^{-i\rho(x)}\{i\frac{d}{dx}(e^{i\rho(x)}f(x)\} \\ &= -i\frac{df}{dx} + \frac{\partial \rho}{\partial x} \\ &= (A_0 + q)f(x) = (Af)(x).\end{aligned}$$

Thus

$$A = WA_0W^{-1} \tag{3.39}$$

and we conclude that A and A_0 are unitarily equivalent with respect to W.

(iii) Spectral properties of A.

Since A and A_0 are unitarily equivalent then A has the same spectral properties as A_0 (see Chapter 4). Specifically, A has no eigenvalues but, rather, a continuous spectrum which is the whole real line.

A diagonalisation result for A can be obtained by noticing that

$$\begin{aligned}(FW^{-1}AF)(k) &= (FW^{-1}[WA_oW^{-1}]f)(k)\\ &= (FA_0W^{-1}f)(k)\\ &= k((W^{-1}f))^{\wedge}(k)\\ &= kF(W^{-1}f)(k) = k(FW^{-1}f)(k).\end{aligned}$$

Thus if we write

$$T := FW^{-1}, \qquad Tf := \tilde{f} \tag{3.40}$$

then we see that T, a "generalised" type of Fourier transform, diagonalises A in the sense that T replaces A acting on f by multiplication by k on $\tilde{f}$

3.5 Comparison of solutions

As we have already mentioned in Chapter 1, an immediate comparison can be obtained by enquiring whether or not solutions to the perturbed problem are asymptotically equal to solutions to the free problem. This will be the case when

$$\lim_{t\to\pm\infty} \| v - u \|_H = 0. \tag{3.41}$$

Using (3.13), (3.35) and the unitarity of U_0 we see that (3.41) can be written

$$\begin{aligned}0 &= \lim_{t\to\pm\infty} \| U(t)g - U_0(t)f \|_H\\ &= \lim_{t\to\pm\infty} \| U_0^*(t)U(t)g - f \|_H\\ &= \| \Omega_{\pm}g - f \|_H \end{aligned}\tag{3.42}$$

where

$$\Omega_{\pm} := \lim_{t\to\pm\infty} U_0^*(t)U(t) =: \lim_{t\to\pm\infty} \Omega(t) \tag{3.43}$$

are the so-called **wave operators** associated with A and A_0. We would remark that in later sections we will find it more convenient to define the wave operators as the **inverse** of those introduced here.

Thus we see that the two systems will be asymptotically equal as $t \to \pm\infty$ provided the initial data for the perturbed and free problems are related according to

$$\Omega_{\pm}g = f_{\pm} \tag{3.44}$$

where the subscripts on f are included to indicate that different initial values may have to be considered for the free problem when investigating $t \to +\infty$ and $t \to -\infty$.

To prove the existence of the limits appearing in (3.41) to (3.43) is, in general, quite an undertaking. We shall discuss this aspect in more detail in a later chapter. However, in the present, particular, case the situation is eased because we can actually compute the various quantities involved. Starting with (3.39) we obtain

$$\begin{aligned} A &= WA_0W^{-1} \\ A^2 &= A = (WA_0W^{-1})(WA_0W^{-1}) = WA_0^2W^{-1} \\ &\text{-------------------------} \\ A^n &= WA_0^nW^{-1}. \end{aligned} \tag{3.45}$$

Furthermore

$$\begin{aligned} U(t) = \exp(-itA) &= (1 + (-itA) + \frac{1}{2!}(-itA)^2 + \cdots) \\ &= W(1 + (-itA_0) + \frac{1}{2!}(-itA_0)^2 + \cdots)W^{-1} \\ &= W\exp(-itA_0)W^{-1} = WU_0(t)W^{-1}. \end{aligned} \tag{3.46}$$

The action of the unitary operator $U(t)$ can now be simply obtained, using (3.29) and (3.38), in the form

$$\begin{aligned} U(t)f(x) &= WU_0(t)W^{-1}f(x) \\ &= e^{-i\rho(x)}U_0(t)e^{i\rho(x)}f(x) \\ &= e^{-i\rho(x)}e^{i\rho(x-t)}f(x-t). \end{aligned} \tag{3.47}$$

We now find, using (3.29) and (3.47), that

$$\begin{aligned} \Omega(t)f(x) :&= U_0^*(t)U(t)f(x) = U_0(-t)U(t)f(x) \\ &= U_0(-t)\{e^{-i\rho(x)}e^{i\rho(x-t)}f(x-t)\} \\ &= e^{-i\rho(x+t)}e^{i\rho(x)}f(x). \end{aligned}$$

Since

$$\rho(x) = \int_0^x q(y)dy \tag{3.37}$$

we have

$$\rho(x) - \rho(x+t) = -\int_x^{x+t} q(y)dy.$$

Consequently

$$\Omega(t) = \exp\{-i\int_x^{x+t} q(y)dy\}f(x). \tag{3.48}$$

Therefore, the wave operators (3.43) are given by

$$\Omega_\pm = \lim_{t\to\pm\infty} \Omega(t) = \exp\{-i\int_x^{\pm\infty} q(y)dy\}. \tag{3.49}$$

In this particular case the wave operators are operators of multiplication by $\exp\{-i\int_x^{\pm\infty} q(y)dy\}$.

Moreover, since

$$\Omega_\pm^{-1} = \exp\{i\int_x^{\pm\infty} q(y)dy\} = \Omega_\pm^*$$

and

$$|\exp(-i\int_x^{\pm\infty} q(y)dy)| = 1$$

it follows that $\Omega_\pm$ are unitary operators on $L_2(\mathbf{R})$.

We also notice that

$$\begin{aligned}(\Omega_\pm A_0 \Omega_\pm^{-1})f(x) &= \exp\{-i\int_x^{\pm\infty} q(y)dy\}\{-i\frac{\partial}{\partial x}[\exp(i\int_x^{\pm\infty} q(y)dy)f(x)]\} \\ &= -i\frac{\partial}{\partial x}f(x) + q(x)f(x) = Af(x). \end{aligned} \tag{3.50}$$

Hence A and A_0 are unitarily equivalent with respect to $\Omega_\pm$.

Furthermore, since F diagonalises A_0,

$$F\Omega_\pm^{-1}Af = F\Omega_\pm^{-1}\{\Omega_\pm A_0\Omega_\pm^{-1}\}f = FA_0\Omega_\pm^{-1}f = kF\Omega_\pm^{-1}f. \tag{3.51}$$

Setting

$$G := F\Omega_\pm^{-1} \tag{3.52}$$

in (3.51) we find that $GAf = kGF$.

Thus G is a generalised type of Fourier Transform which diagonalises A in a similar manner to which F diagonalised A_0.

Collecting these various results we find that the relation (3.44) can be written, in this particular case, as

$$f_{\pm}(x) = \exp\{-i\int_{x}^{\pm\infty} q(y)dy\}g(x). \tag{3.53}$$

Thus we have determined the initial data for the unperturbed problem, in terms of the given data for the perturbed problem, which is required to ensure that the solutions to the two problems are asymptotically equal as $t \to \pm\infty$.

We would remark that in this particular case we can actually solve the perturbed problem (3.33), (3.34). The problem can be reduced, using the transform T defined in (3.40) to read

$$\tilde{v}_t + ik\tilde{v} = 0, \quad \tilde{v}(0) = \tilde{g}. \tag{3.54}$$

This has the solution

$$\tilde{v} = e^{-ikt}\tilde{g}$$

from this we can obtain

$$\begin{aligned} v(x,t) &= (T^{-1}\tilde{v})(x,t) = (FW^{-1})^{-1}\{e^{-ikt}\tilde{g}\}(x) \\ &= WF^{-1}\{e^{-ikt}(FW^{-1}g)\}x) = WF^{-1}\{e^{-ikt}Fh\}(x) \\ &= WF^{-1}\{e^{-ikt}\hat{h}\}(x) \\ &= Wh(x-t). \end{aligned}$$

Therefore

$$v(x,t) = \exp\{-i\rho(x)\}\exp\{i\rho(x-t)\}g(x-t). \tag{3.55}$$

Direct substitution in (3.33), (3.34) readily indicates that (3.55) is the required solution. However, we would emphasise that there is no guarantee that the solution (3.13) and (3.55) become close as $t \to \pm\infty$; this can only be ensured when the requirement (3.44) is satisfied.

3.6 Summary

In this Chapter we have considered a specific unperturbed (easy) problem and an associated perturbed (hard) problem. These problems were characterised by a linear, self-adjoint operator A_0 and A respectively. We have shown that these operators are unitarily equivalent with respect to a certain operator W which we introduced. Hence we were able to infer that A and A_0 had the same spectral properties. As we assume that the spectral properties of A_0 are easily obtainable this result is of considerable advantage. We have also obtained operators which diagonalise A_0 and A. This result will prove to be particularly useful when we come to discuss spectral expansions associated with operators such as A_0 and A.

However, promising as this strategy appears to be, it does rely very heavily on being able to **guess** an appropriate form for W. Furthermore if our attention is confined entirely to a development based on a knowledge of W then there is no obvious mechanism immediately available which will enable us to compare the solutions of the easy and hard problems as $t \to \pm\infty$. For more general problems than that considered here a more systematic approach is required. A possible such approach has already been hinted at in this Chapter and is one which relies on the existence and on the various properties of the wave operators associated with the operators A_0 and A. Specifically, for more general problems we shall develop a strategy along the following lines.

(i) Obtain representations of the easy and hard problems which are characterised by the self-adjoint operator A_0 and A respectively.

(ii) Determine the spectral properties of A_0.

(iii) Prove the existence of wave operators associated with A_0 and A.

(iv) Prove the existence of suitable initial data for the easy problem which will ensure the asymptotic equality of the easy and hard problem as $t \to \pm\infty$. This will be settled by establishing the so-called completeness of the wave operators.

(v) Determine the wave operators in terms of spectral expansions associated with A_0 and A.

We shall then use this strategy to discuss two main classes of problems.

(a) Target scattering problems in which the domain of A_0 is perturbed in a manner which characterises the target.

(b) Non-linear scattering problems in which A_0 is perturbed by a potential which depends on the required solution.

Chapter 4
Elements of spectral theory

4.1. Introduction

Spectral Theory is concerned with the representation of often quite complicated operators in terms of simpler operators with properties which are easily understood. This is achieved in a Hilbert space setting in terms of projection operators defined on the spectrum of the given operator.

In finite dimensional spaces linear operators can be represented in terms of matrices with respect to a chosen basis. Under suitable circumstances the matrix concerned can be diagonalised the entries being the eigenvalues of the matrix which in turn are the eigenvalues of the original operator.

The corresponding analogue in infinite dimensional spaces is given by the so-called Spectral Theorem. However, for operators on an infinite dimensional space a major difficulty arises from the fact that their spectrum consists of more than just eigenvalues.

In this chapter we introduce the spectral theorem and explain some of its consequences.

4.2 Basic concepts

Let X be a complex normed linear space and

$$A : X \to X, \quad D(A) \subseteq X$$

a linear operator. With A we associate the operator

$$(A - \lambda I) : X \to X, \quad D(A - \lambda I) = D(A)$$

where $\lambda \in \mathbf{C}$ and I is the identity operator.

If $(A - \lambda I)$ is invertible then we denote

$$R_\lambda(A) \equiv R_\lambda = (A - \lambda I)^{-1} \tag{4.1}$$

and refer to $R_\lambda(A)$ as the resolvent (operator) of A. We notice that as A is a linear then so is $R_\lambda(A)$ and would remark that an investigation of the properties of $R_\lambda(A)$ will be central to a study of the operator A.

In developing spectral theory on infinite dimensional spaces we shall need the following concepts.

Definition 4.1

Let X be a normed linear space and

$$A : X \to X, \quad D(A) \subseteq X$$

a linear operator.

A **regular value** of A is a complex number λ such that

(i) $R_\lambda(A)$ exists

(ii) $R_\lambda(A)$ is bounded

(iii) $R_\lambda(A)$ is defined on a dense subset of X.

The set of all regular values of A, denoted $\rho(A)$, is called the **resolvent set** of A.

Definition 4.2

Let A be the linear operator defined in Definition 4.1.

The **spectrum** of A, denoted $\sigma(A)$, is the complement in the complex plane of $\rho(A)$, that is

$$\sigma(A) = \mathbf{C} \setminus \rho(A). \tag{4.2}$$

The spectrum of A is partitioned into the following three disjoint sets.

(i) The **point spectrum** $\sigma_p(A)$

$\sigma_p(A)$ consists of all those $\lambda \in \mathbf{C}$ such that $R_\lambda(A)$ does not exist.

$\lambda \in \sigma_p(A)$ is called an eigenvalue of A.

(ii) The **continuous spectrum** $\sigma_c(A)$

$\sigma_c(A)$ consists of all those $\lambda \in \mathbf{C}$ such that $R_\lambda(A)$ exists as an unbounded operator and is defined on a dense subset of X.

(iii) The **residual spectrum** $\sigma_R(A)$

$\sigma_R(A)$ consists of all those $\lambda \in \mathbf{C}$ such that $R_\lambda(A)$ exists as either a bounded or an unbounded operator but in either case is not defined on a dense subset of X.

The **spectrum** of A is the union of these disjoint sets

$$\sigma(A) = \sigma_p(A) \cup \sigma_c(A) \cup \sigma_R(A) \tag{4.3}$$

and a $\lambda \in \sigma(A)$ is referred to as a **spectral value** of A.

We notice

$$\mathbf{C} = \rho(A) \cup \sigma(A). \tag{4.4}$$

We see that the spectrum of a linear operator on a finite dimensional space consists only of the point spectrum, that is both the continuous and residual spectra of the operator are empty. In this case the operator is said to have a pure point spectrum and all its spectral values are eigenvalues.

Definition 4.3

Let A be the linear operator defined in Definition 4.1. If $(A - \lambda I)u = 0$ for some non-trivial $u \in D(A)$ then u is an **eigenvector** associated with the **eigenvalue** $\lambda \in \sigma_p(A)$.

The subspace $M_\lambda \subset X$ consisting of the zero element and all eigenvectors of A corresponding to the eigenvalues λ is called the **eigenspace** of A corresponding to the eigenvalue λ.

We now claim that M_λ is a subspace of H whenever A is either bounded or unbounded but closed. That it is a linear manifold is clear since for any $f_1, f_2 \in M_\lambda$ and $\alpha, \beta \in \mathbf{R}$ or $\mathbf{C}$ it is an easy matter to show that

$$A(\alpha f_1 + \beta f_2) = \lambda(\alpha f_1 + \beta f_2).$$

To show that it is closed let $\{f_n\} \subset M_\lambda$ be a sequence such that $f_n \to f$ as $n \to \infty$ and consider the two cases.

(i) A bounded.

A bounded implies A continuous and we have

$$Af = A \lim f_n = \lim Af_n = \lim \lambda f_n = \lambda f$$

thus $f \in M_\lambda$ and we conclude that M_λ is closed.

(ii) A unbounded but closed.

In this case, since $f_n \to f$, there will exist g such that

$$g := \lim Af_n = \lim \lambda f_n = \lambda f.$$

However, A is closed which implies that $f \in D(A)$ and $Af = g$. Hence $f \in M_\lambda$ and again we can conclude that M_λ is closed.

Hence M_λ is a subspace of H.

4.3 Eigenvalues and eigenvectors

For the remainder of this Chapter we consider linear operators defined on a complex, separable Hilbert space, H.

Our main interest is in self-adjoint and unitary operators. The most important properties of their eigenvalues and eigenvectors are contained in the following.

Theorem 4.4

The eigenvalues of a self-adjoint operator are real.

Proof: Let $A : H \to H$ be self-adjoint and $\lambda \in \sigma_p(A)$ with associated eigenvector u, then

$$\lambda(u, u) = (\lambda u, u) = (Au, u) = (u, Au) = \bar{\lambda}(u, u)$$

which, because u is non-trivial, implies $\lambda = \bar{\lambda}$. Hence $\sigma_p(A) \subset \mathbf{R}$. ∎

Theorem 4.5

The eigenvalues of a unitary operator are complex numbers of modulus one.

Proof: Let $U : H \to H$ be a unitary operator and $\mu \in \sigma_p(U)$ with associated eigenvector w, then

$$(w, w) = (Uw, Uw) = (\mu w, \mu w) = \bar{\mu}\mu(w, w)$$

which implies $|\mu|^2 = 1$ because w is non-trivial. ∎

Theorem 4.6

The eigenvectors of either a self-adjoint or unitary operator corresponding to different eigenvalues are orthogonal.

Proof: Let $A : H \to H$ be self-adjoint and different, $\lambda_1, \lambda_2 \in \sigma_p(A)$ with associated eigenvectors u_1, u_2 then

$$\begin{aligned}(\lambda_1 - \lambda_2)(u_1, u_2) &= (\lambda_1 u_1, u_2) - (u_1, \lambda_2 u_2)\\ &= (Au_1, u_2) - (u_1, Au_2)\\ &= (Au_1, u_2) - (Au_1, u_2) = 0\end{aligned}$$

which implies $(u_1, u_2) = 0$ since $\lambda_1 \neq \lambda_2$.

Let $U : H \to H$ be unitary and different $\mu_1, \mu_2 \in \sigma_p(U)$ with associated eigenvectors w_1, w_2 then

$$\mu_1 \bar{\mu}_2 (w_1, w_2) = (\mu_1 w_1, \mu_2 w_2) = (Uw_1, Uw_2) = (w_1, w_2)$$

which implies that $(w_1, w_2) = 0$ when $\mu_1 \bar{\mu}_2 \neq 1$. If $\mu_1 \neq \mu_2$ then $\mu_1 \bar{\mu}_2 \neq 1$ because if $\mu_1 \bar{\mu}_2 = 1$ then

$$\mu_1 = \mu_1 \mid \mu_2 \mid = \mu_1 \bar{\mu}_2 \mu_2 = \mu_2$$

which is a contradiction. ∎

Theorem 4.7

Let H be a complex, separable Hilbert space and $A, B : H \to H$ linear operators. If B is bounded and B^{-1} exists then A and BAB^{-1} have the same eigenvalues.

Proof: If λ is an eigenvalue of A then there exists a non-trivial ϕ such that

$$A\phi = \lambda\phi.$$

If B^{-1} exists then $B\phi$ is not zero for all non-trivial ϕ and

$$BAB^{-1}B\phi = BA\phi = B\lambda\phi = \lambda B\phi$$

which implies that λ is also an eigenvalue of the operator BAB^{-1}; the corresponding eigenvector is $B\phi$. Conversely, let μ be an eigenvalue of BAB^{-1}, then there exists a non-trivial ψ such that

$$BAB^{-1}\psi = \mu\psi.$$

Consequently,

$$\begin{aligned}B^{-1}BAB^{-1}\psi &= \mu B^{-1}\psi\\ AB^{-1}\psi &= \mu B^{-1}\psi\end{aligned}$$

which implies that μ is an eigenvalue of A with associated eigenvector $B^{-1}\psi$. ∎

Let λ be an eigenvalue of the linear operator $A : H \to H$ and let

$$M_\lambda = \{\psi \in D(A) \subset H : A\psi = \lambda\psi\}$$

denote the eigenspace of A corresponding to the eigenvalue λ. On this subspace A acts like the scalar multiple λ of the identity operator.

For a self-adjoint or unitary operator the eigenspaces corresponding to different eigenvalues are orthogonal and the operator acts on their direct sum like a diagonal matrix. To see this latter point let A be a self-adjoint or unitary operator and let $\lambda_1, \lambda_2, \cdots, \lambda_k, \cdots$ denote its different eigenvalues. For each k we write M_k for the eigenspace corresponding to λ_k. An orthonormal basis for M_k is denoted $\{\phi_s^k\}_s$. It is worth remarking that whilst the number of eigenvalues may be either finite or infinite they are always countable. If this were not so then there would be an uncountable number of different eigenvalues with an associated uncountable number of orthonormal basis vectors. This is impossible in a separable Hilbert space.

The dimension of M_k, which is the number of basis elements ϕ_s^k for M_k, may be finite or infinite and may be different for different values of k.

Since eigenvectors for different eigenvalues of A are orthogonal the set of all vectors ϕ_s^k for different k is orthonormal, that is

$$(\phi_r^k, \phi_s^m) = \delta_{rs}\delta_{km}.$$

Consider the subspace

$$H_p = \bigoplus_k M_k$$

which consists of all linear combinations of the form

$$\sum_{k,s} a_s^k \phi_s^k.$$

We shall refer to this as the **point subspace** of A; it is the subspace spanned by all eigenvectors of A. Evidently the set of vectors ϕ_s^k, $k, s = 1, 2, \cdots$ is an orthonormal basis for H_p. Thus we have

$$A\phi_s^k = \lambda_k \phi_s^k$$

and

$$(\phi_r^j, A\phi_s^k) = \lambda_k \delta_{jk} \delta_{rs}.$$

Thus on H_p the operator A acts like a diagonal matrix. Its off diagonal elements are zero whilst its diagonal elements are eigenvalues of A. For any $\psi \in H_p$ there are scalars a_s^k such that

$$\psi = \sum_{k,s} a_s^k \phi_s^k \tag{4.5}$$

and

$$A\psi = \sum \lambda_k a_s^k \phi_s^k. \tag{4.6}$$

Equations (4.5), (4.6) provide an example of a **spectral representation** for the operator A.

Another concept which will be of use later is that of a reducing subspace. We shall introduce this in terms of bounded operators. For unbounded operators the following definition and two theorems require a rather more careful statement which properly takes account of the domains involved.

Definition 4.8

A closed subspace M **reduces** a linear operator A if

(i) $A\psi \in M$ for every $\psi \in M$.

(ii) $A\phi \in M^{\perp}$ for every $\phi \in M^{\perp}$.

Theorem 4.9

Let

(i) H be a complex separable Hilbert space

(ii) $A : H \to H,\ D(A) \subseteq H$ a linear operator

(iii) $E : H \to M$, a projection operator onto the subspace M.

Then the following statements are equivalent

(a) M reduces A

(b) $EA = AE$

(c) $(I - E)A = A(I - E)$.

Proof: It is obvious that (b) and (c) are equivalent. We shall show that (a) $\Rightarrow$ (b) and that ((b), (c)) $\Rightarrow$ (a).

For any element $\psi \in H$, we have by the Projection Theorem

$$\psi = u + v, \quad u \in M, \quad v \in M^{\perp}.$$

Hence

$$A\psi = Au + Av.$$

(a) $\Rightarrow$ (b):

M reduces $A \Rightarrow Au \in M$ and $Av \in M^{\perp}$.

Therefore,

$$EA\psi = Au = AE\psi \quad \text{for all } \psi \in H.$$

Hence (a) $\Rightarrow$ (b).

((b), (c)) $\Rightarrow$ (a):

If $u \in M$ and $EA = AE$ then $Eu = u$ and

$$Au = AEu = EAu.$$

Thus $Au \in M$ for all $u \in M$. If $v \in M^{\perp}$ and $(I - E)A = A(I - E)$ then $(I - E)v = v$ and

$$Av = A(I - E)v = (I - E)Av.$$

Thus $Av \in M^{\perp}$ for all $v \in M^{\perp}$.

Hence ((b), (c)) $\Rightarrow$ (a). ■

This theorem indicates that if M reduces A then

$$A = AE + A(I - E) \tag{4.7}$$

which implies that A is the sum of two parts, namely AE an operator on M and $A(I - E)$ an operator on $M^{\perp}$.

Theorem 4.10

Let A be a bounded, linear operator which is either self-adjoint or unitary on a complex separable Hilbert space H. Let H_p be the subspace spanned by the eigenvectors of A then H_p reduces A. The operators induced by A in H_p and $H_p^\perp$ are again self-adjoint and unitary respectively.

Proof (i): Assume that A is self-adjoint.

Let $u \in H_p$, then (4.5), (4.6) indicate that $Au \in H_p$. If $v \in H_p^\perp$ then

$$(u, Av) = (Au, v) = 0 \quad \text{for any } u \in H_p.$$

Hence $Av \in H_p^\perp$ and we can conclude that H_p reduces A.

Let $P : H \to H_p$ be a projection. Then there exist operators A_1 in H_p and A_2 in $H_p^\perp$ with domains

$$D(A_1) = PH = H_p$$
$$D(A_2) = (I - P)H = P^\perp H = H_p^\perp$$

such that

$$Af = A_1(Pf) + A_2(P^\perp f), \quad f \in H.$$

The operators A_1, and A_2 are the operators **induced** by A in H_p and $H_p^\perp$ respectively.

For $f, g \in H_p$ we have, recognising this last relation,

$$(A_1 f, g) = (APf, g) = (PAf, g) \qquad \text{(by Theorem 4.9)}$$
$$= (f, APg) = (f, Ag) = (f, A, g).$$

Thus A_1 is a self adjoint operator in H_p. Similarly, A_2 is self adjoint in $H_p^\perp$.

(ii): Assume that A is unitary.

As before $u \in H_p$ implies $Au \in H_p$.

Let $v \in H_p^\perp$ and let w be an eigenvector of A with associated eigenvalue λ, then

$$Aw = \lambda w$$

and

$$\lambda(w, Av) = (Aw, Av) = (w, v) = 0$$

which implies $(w, Av) = 0$ since $|\lambda| = 1$. Thus, Av is orthogonal to every vector in the orthonormal basis (of eigenvectors of A) for H_p. Therefore $Av \in H_p^{\perp}$ and hence H_p reduces A.

With A_1, A_2 defined as in (i) we have for A unitary

$$\| f \| = \| Af \| = \| A_1 f \| \quad \forall f \in H_p.$$

Therefore A_1 is an isometric linear operator on H_p. Similarly, A_2 is an isometric linear operator on $H_p^{\perp}$. To show that A_1 is unitary it remains to show that A_1 maps H_p onto H_p. Let $g \in H_p$ be given. Since A is unitary then A maps H onto H and there exists an element $f \in H$ such that

$$Af = g.$$

Since $Af \in H_p$ and $(A_1 Pf) \in H_p$, $(A_2 P^{\perp} f) \in H_p^{\perp}$ we conclude from the decomposition of Af given in (i) that $A_2 P^{\perp} f = 0$ and that $\| P^{\perp} f \| = \| A_2 P^{\perp} f \| = 0$. Consequently $f \in H_p$ and

$$g = Af = A_1 f \in H_p.$$

■

In future, we shall refer to H_p as the **point subspace** of A. Furthermore $H_p^{\perp}$ the orthogonal complement of H_p, will be denoted by H_c and referred to as the **subspace of continuity** of A.

Thus we see from (4.7) that a self-adjoint or unitary operator splits into two parts. One part is an operator acting on the closed subspace spanned by the eigenvectors and as such can be represented by a diagonal matrix of eigenvalues with respect to an orthonormal basis of eigenvectors. The other part is an operator acting on the orthogonal complement of the subspace spanned by the eigenvectors. Again we would emphasise that we are assuming that our linear operators are either bounded or unbounded but closed in order that we can work with subspaces.

The simplest possibility in this decomposition is that the eigenvectors of the operator under consideration span the whole space. In this case the operator corresponds to a diagonal matrix with respect to an orthonormal basis of eigenvectors which span the space.

Example 4.11

Let $P : H \to N$ be a projection operator onto the subspace N. Assume there exists a non-trivial $\psi \in H$ such that

$$P\psi = \lambda\psi, \quad \lambda \in \mathbf{C}.$$

Then

$$\lambda^2\psi = \lambda P\psi = P^2\psi = P\psi = \lambda\psi$$

which implies that

$$\lambda = 1 \text{ or } 0. \tag{4.8}$$

We therefore conclude that a projection operator can have only the two different eigenvalues indicated in (4.8).

Furthermore, if $\psi \in N$ then $P\psi = \psi$ by definition of the projection operator P. Also if $\phi \in N^{\perp}$ then $P\phi = 0$. Therefore, recognising the facts that a projection operator is self-adjoint and that eigenspaces of self-adjoint operators corresponding to different eigenvalues are orthogonal we infer that

$$N = \text{eigenspace of } A \text{ corresponding to the eigenvalue } \lambda = 1$$

$$N^{\perp} = \text{eigenspace of } A \text{ corresponding to the eigenvalue } \lambda = 0.$$

A basis for the whole space is obtained by combining the basis for N and the basis for $N^{\perp}$. The projection operator P can be represented as a diagonal matrix with respect to this basis; the diagonal elements are one for basis vectors from N and zero for basis vectors from $N^{\perp}$.

Thus we have completely described a projection operator in terms of its eigenvalues and eigenvectors.

4.4 Spectral decompositions on finite dimensional spaces

On finite dimensional spaces the only spectral values are eigenvalues. However, in the associated analysis it is necessary to work with complex scalars as there are linear operators on real finite-dimensional spaces that have no eigenvalues.

Example 4.12

The operator A represented by the matrix

$$\begin{bmatrix} 0 & 1 \\ -1 & 0 \end{bmatrix}$$

has eigenvalues $\pm i$.

For this reason we shall always work with complex separable Hilbert spaces.

Theorem 4.13

Let A be a linear operator on a finite dimensional space. Equivalent necessary and sufficient conditions for λ to be an eigenvalue of A are

(i) The operator $(A - \lambda I)$ has no inverse.

(ii) The determinant of the matrix corresponding to $(A - \lambda I)$ is zero.

Proof (i): The operator $(A - \lambda I)$ has an inverse if and only if $(A - \lambda I)\phi = 0$ implies $\phi = 0$.

(ii) The operator $(A - \lambda I)$ has an inverse if and only if $\mathcal{A}$ the matrix corresponding to $(A - \lambda I)$ has an inverse. That is, if and only if the determinant of the matrix $\mathcal{A}$ is non zero. ∎

The above Theorem indicates a means of calculating the eigenvalues of the operator A. In an n-dimensional space the operator $(A - \lambda I)$ can be represented in terms of an $(n \times n)$-matrix. If we set the determinant of this matrix to zero then we obtain an n-th order polynomial equation to solve for the eigenvalues λ. Such equations always have at least one solution and never more than n distinct solutions. Consequently, a linear operator on a complex n-dimensional space always has at least one eigenvalue and at most n distinct eigenvalues.

Theorem 4.14

On a finite-dimensional complex space the eigenvectors of a self-adjoint or of a unitary operator span the whole space.

Proof: Let A be a linear self-adjoint or unitary operator on a finite-dimensional complex space, H, and let M denote the subspace spanned by its eigenvectors. Let $E : H \to M$ be a projection operator onto M. Theorem 4.10 indicates that M reduces A.

Suppose $M \neq H$ and consider the operator $A(I - E)$ on $M^{\perp}$. Since every operator on a finite-dimensional space has at least one eigenvalue there must exist a scalar λ and a non-trivial element $v \in M^{\perp}$ such that

$$A(I - E)v = \lambda v.$$

Consequently

$$Av = AEv + A(I - E)v = A(I - E)v = \lambda v$$

which implies that v is an eigenvalue of A, which contradicts the assumption that M is not the whole space. This establishes the Theorem. ∎

The above Theorem means that for every self-adjoint or unitary operator A on an n-dimensional complex space there is a basis for the space consisting entirely of orthonormal eigenvectors of A. Let

(i) $\lambda_1, \cdots, \lambda_m$ be the the different eigenvalues of A.

(ii) M_k the eigenspace corresponding to the eigenvalue λ_k, $k = 1, 2, \cdots, m$.

(iii) ϕ_s^k be an orthonormal basis for M_k, $k = 1, \cdots m$.

Then

$$A\phi_s^k = \lambda_k \phi_s^k, \ k = 1, 2, \cdots m, \ s = 1, 2, \cdots, s(k)$$

where $s(k)$ is a positive integer depending on k, and $\{\phi_s^k\}_{k,s}$ is an orthonormal basis for the whole space. The total number of eigenvectors ϕ_s^k is n and $m \leq n$. Also the number of orthonormal eigenvectors ϕ_s^k associated with the eigenvalue λ_k may be different for each k.

In the case when A has precisely n distinct eigenvalues then $m = n$ and the subscript s is unnecessary since there is only one eigenvector associated with any λ_k.

For convenience we collect together some results we have obtained previously concerning the basis elements ϕ_s^k.

Orthogonality:

$$(\phi_s^k, \phi_r^\ell) = \delta_{k\ell}\delta_{rs} \tag{4.9}$$

Spectral Representation:

For any element $\psi \in H$ we have seen that there are scalars a_s^k such that

$$\psi = \sum_{k,s} a_s^k \phi_s^k \tag{4.5}$$

$$A\psi = \sum_{k,s} a_s^k \lambda_k \phi_s^k. \tag{4.6}$$

Taking the inner product of (4.5) with ϕ_r^ℓ and using (4.9) determines the scalars a_r^m in the form

$$a_r^m = (\psi, \phi_r), \quad \ell = 1, 2 \cdots m, \quad s = 1, 2, \cdots s(\ell). \tag{4.10}$$

A closer look at (4.5) suggests that we define the operator

$$P_k : H \to M_k, \quad k = 1, 2, \cdots, m \tag{4.11}$$

$$\psi \to \sum_{s=1}^{s(k)} a_s^k \phi_s^k. \tag{4.12}$$

The representations (4.5) can now be written

$$\psi = \sum_{k=1}^{m} P_k \psi \tag{4.13}$$

which implies the completeness property

$$\sum_{k=1}^{m} P_k = I. \tag{4.14}$$

The operator P_k is a projection onto the eigenspace M_k. Since eigenspaces corresponding to different eigenvalues of a self-adjoint or unitary operator are orthogonal it follows that the projections P_k, $k = 1, 2, \cdots, m$ are orthogonal in the sense that

$$P_k P_\ell = \delta_{k\ell} P_\ell \tag{4.15}$$

Furthermore, (4.5) can now be written

$$A\psi = \sum_{k,s} a_s^k \lambda_k \phi_s^k = \sum_k \lambda_k P_k \psi$$

which implies

$$A = \sum_{k=1}^{m} \lambda_k P_k. \tag{4.16}$$

This is a representation of the operator A in terms of projections. It illustrates how the spectrum of A can be used to provide a representation of A in terms of simpler operators. Whilst this use of projections seems quite natural, unfortunately the idea does not generalise immediately to infinite-dimensional spaces. We now describe a slightly different approach for obtaining a representation of A which has the advantage that it does generalise to infinite dimensional spaces where the spectrum of an operator can be quite complicated.

First we consider a self-adjoint A and order its distinct eigenvalues in the form

$$\lambda_1 < \lambda_2 < \cdots < \lambda_{m-1} < \lambda_m.$$

For each $\lambda \in \mathbf{R}$ we define

$$E_\lambda = \sum_{\lambda_k \leq \lambda} P_k. \tag{4.17}$$

Clearly, E_λ is a projection operator onto the subspace spanned by all eigenvectors associated with eigenvalues $\lambda_k < \lambda$.

The projections E_λ have the properties

(a) $E_\lambda = 0 \quad \lambda < \lambda_1$.

(b) $E_\lambda = I \quad \lambda \geq \lambda_m$.

This follows from (4.14). Furthermore from (4.17)

(c) $E_\mu E_\lambda = E_\lambda E_\mu = E_\mu, \quad \mu \leq \lambda$.

When (c) holds we write

$$E_\mu \leq E_\lambda, \quad \mu \leq \lambda \tag{4.18}$$

These properties indicate that E_λ changes from the zero operator to the identity operator as λ runs through the spectrum (eigenvalues) of A. Furthermore, we notice that E_λ changes by P_k when λ reaches λ_k.

For each λ define

$$dE_\lambda = E_\lambda - E_{\lambda-\epsilon}, \quad \epsilon > 0. \tag{4.19}$$

If $\epsilon > 0$ is small enough to ensure that there is **no** λ_k such that $\lambda - \epsilon < \lambda_k < \lambda$ then $dE_\lambda = 0$.

If $\lambda = \lambda_k$ then $dE_\lambda = dE_{\lambda_k} = P_k$.

We now recall the definition of the Riemann-Stieltjes integral of a function g with respect to a function f

$$\int_a^b g(x)df(x) = \lim_{n\to\infty} \sum_{j=1}^{n} g(x_j)[f(x_j) - f(x_{j-1})] \tag{4.20}$$

where $a = x_0 < x_1 < \cdots < a_n = b$ is a partition of the range of integration.

With the above definition in mind we have

$$\int_{-\infty}^{\infty} dE_\lambda = \lim_{n \to \infty} \sum_{j=1}^{n} 1(E_{x_j} - E_{x_{j-1}})$$

where $\lambda_1 \le x_0 < x_1 < \cdots < x_n \le \lambda_m$. Combining this with (4.14) we obtain

$$\int_{-\infty}^{\infty} dE_\lambda = I. \tag{4.21}$$

Furthermore starting with (4.16) and arguing as above we have

$$A = \sum_{k=1}^{m} \lambda_k P_k = \lim_{n \to \infty} \sum_{j=1}^{n} \lambda_{x_j} (E_{x_j} - E_{x_{j-1}})$$

which implies

$$A = \int_{-\infty}^{\infty} \lambda dE_\lambda. \tag{4.22}$$

This is the spectral representation of the self-adjoint operator A with eigenvalues $\lambda_1 < \lambda_2 < \cdots < \lambda_m$ on an n-dimensional complex Hilbert space.

For arbitrary elements ψ and ϕ in the n-dimensional space the above result lead to

$$(\psi, \phi) = \int_{-\infty}^{\infty} d(E_\lambda \psi, \phi) = \int_{-\infty}^{\infty} dw(\lambda) \tag{4.23}$$

$$(A\psi, \phi) = \int_{-\infty}^{\infty} \lambda d(E_\lambda \psi, \phi) = \int_{-\infty}^{\infty} \lambda dw(\lambda). \tag{4.24}$$

Here $w(\lambda) := (E_\lambda \psi, \phi)$ defines a complex valued function of λ which changes by $(P_k \psi, \phi)$ at $\lambda = \lambda_k$.

For a unitary operator U with eigenvalues $\lambda_k = e_k^{i\theta_k}$ ordered in the form

$$0 < \theta_1 < \theta_2 < \cdots < \theta_m \le 2\pi$$

we argue in a similar manner and obtain the spectral representation

$$U = \int_0^{2\pi} e^{i\lambda} dE_\lambda. \tag{4.25}$$

This leads as before to the expression

$$(\psi, \phi) = \int_0^{2\pi} e^{i\lambda} d(E_\lambda \psi, \phi). \tag{4.26}$$

Finally, in this section we notice that the projection operators E_λ, (considered as functions of λ) are continuous from the right but discontinuous from the left, that is

$$E_{\lambda+\epsilon}\psi \to E_\lambda \psi, \quad \epsilon \to 0 \tag{4.27}$$

$$dE_\lambda = E_\lambda - E_{\lambda-\epsilon} = P_k \quad \text{as } \lambda \to \lambda_k. \tag{4.28}$$

4.5 Spectral decompositions on infinite-dimensional spaces

On infinite-dimensional spaces matters are more complicated; there are self-adjoint and unitary operators which have no eigenvalues and the spectrum can consist of more than just the point spectrum. Nevertheless, it is still possible to obtain spectral decomposition in terms of projection operators which have similar integral forms to those developed in the previous section.

Definition 4.15

A family of projection operators E_λ depending on the parameter λ is said to be a **spectral family** if it has the properties:

(i) If $\mu \leq \lambda$ then $E_\mu \leq E_\lambda$, that is

$$E_\lambda E_\mu = E_\mu E_\lambda = E_\mu.$$

(ii) If $\epsilon > 0$ then for any element ψ and scalar λ

$$\begin{aligned} E_{\lambda+\epsilon}\psi &\to E_\lambda\psi \quad \text{as } \epsilon \to 0 \\ E_\lambda\psi &\to 0 \quad \text{as } \lambda \to -\infty \\ E_\lambda\psi &\to \psi \quad \text{as } \lambda \to +\infty. \end{aligned}$$

The following Theorem can now be established.

Theorem 4.16 (Spectral Theorem)

Let H be a complex, separable Hilbert space.

(i) For each self-adjoint, linear operator T on H there exists a unique spectral family $\{E_\lambda\}$ such that

$$(T\psi,\phi) = \int_{-\infty}^{\infty} \lambda d(E_\lambda\psi,\phi) \tag{4.29}$$

for all $\psi,\phi \in D(T) \subseteq H$.

(ii) For each unitary, linear operator U on H there exists a unique spectral family $\{E_\lambda\}$ such that $E_\lambda = 0$ for $\lambda \leq 0$ and $E_\lambda = I$ for $\lambda \geq 2\pi$ such that

$$(U\psi,\phi) = \int_{0}^{2\pi} e^{i\lambda} d(E_\lambda\psi,\phi) \tag{4.30}$$

for all $\psi,\phi \in D(U) \subseteq H$.

We write (4.29), (4.30) in the more convenient forms

$$T = \int_{-\infty}^{\infty} \lambda dE_\lambda \tag{4.31}$$

$$U = \int_{0}^{2\pi} e^{i\lambda} dE_\lambda \tag{4.32}$$

and refer to them as the **spectral decompositions** of T and U respectively.

The proof of this Theorem, which is straightforward but quite lengthy, is given in the standard texts mentioned in the Commentary.

Example 4.17

As the complex Hilbert space take $H = L_2(0,1)$. Define $T : H \to H$ by

$$(T\psi)(x) = x\psi(x), \quad \psi \in D(T) = H.$$

Clearly T is linear, self-adjoint.

Let $\{E_x\}$ be the family of projection operators defined by

$$(E_x\psi)(z) = \begin{cases} \psi(z) & z \leq x \\ 0 & z > x \end{cases}$$

The following are immediate.

(i) $\quad E_xE_y = E_yE_x = E_x, \quad x \leq y$

that is $E_x \leq E_y, \quad x \leq y.$

(ii) $\| E_{x+\epsilon}\psi - E_x\psi \|^2 = \int_x^{x+\epsilon} | \psi(y) |^2 \, dy \quad \to 0 \text{ as } \epsilon \to 0.$

Thus $E_{x+\epsilon} \to E_x$ as $E \to 0$.

Since it is reasonable to assume that $\psi(x)$ is zero outside the interval [0,1] then it follows that

$$E_x = 0, \quad x < 0, \; E_x = I, \quad x > 1.$$

Consequently the family $\{E_x\}$ is a spectral family.

A spectral decomposition for T is obtained by noticing

$$\begin{aligned}
\int_{-\infty}^{\infty} x d(E_x\psi, \phi) &= \int_{-\infty}^{\infty} x d \left[\int_0^1 E_x\psi(y)\overline{\phi(y)} dy \right] \\
&= \int_0^1 x d \left[\int_0^x \psi(y)\overline{\phi(y)} dy \right] \\
&= \int_0^1 x\psi(x)\overline{\phi(x)} dx \\
&= (T\psi, \phi).
\end{aligned}$$

The limits follow from the definition of E_x and the fact that E_x is a constant function of x outside the interval [0,1].

Thus T has the spectral representation

$$T = \int_{-\infty}^{\infty} x dE_x. \tag{4.33}$$

We also notice that

$$(E_x\psi, \phi) - (E_{x-\epsilon}\psi, \phi) = \int_{x-\epsilon}^{x} \psi(y)\overline{\phi(y)} dy$$

the right-hand side tending to zero as $\epsilon \to 0$. Therefore $(E_x\phi, \psi)$, as a functional of x, is also continuous from the left. Thus the function w defined by

$$w(x) = (E_x\psi, \phi)$$

is a continuous function of x.

Example 4.18

Define $\quad T : H \to H = L_2(\mathbf{R})$

$$(T\psi)(x) = x\psi(x), \quad \psi \in D(T) = H.$$

With a spectral family defined as in Example 4.17 it is readily shown that T also has the spectral decomposition (4.33). However the spectral family in this case has the properties that E_x increases over the whole range $-\infty < x < \infty$ with $E_x \to 0$ as $x \to -\infty$ and $E_x \to I$ as $x \to +\infty$.

These two examples illustrate some of the difficulties which arise when working in infinite-dimensional spaces. We see that E_λ, as a function of λ, may increase in a continuous manner rather than by discrete jumps. In these examples E_λ is continuous for all λ; there are no jumps in its value. This is because the operators T in each case had no eigenvalues.

4.6 Properties of spectral families

The spectral family $\{E_\lambda\}$ associated with a self-adjoint operator T on a Hilbert space H provides information regarding the spectrum of T in quite a simple manner. We have already seen in §4.4 that $\{E_\lambda\}$ has a discontinuous behaviour at the eigenvalues of the associated operator (see (4.19)). This property as we shall see carries over to an infinite dimensional setting. Furthermore, information about the other parts of the spectrum which can exist in an infinite dimensional setting can also be obtained. We begin by examining the point spectrum.

Theorem 4.19

Let $T : H \to H$ be a linear, self-adjoint operator with an associated spectral family $\{E_\lambda\}$ and spectral decomposition

$$T = \int_{-\infty}^{\infty} \lambda dE_\lambda.$$

Then E_λ has a discontinuity at $\lambda = \mu$ if and only if μ is an eigenvalue of T.

Let $P_\mu : H \to M_\mu$ denote the projection operator onto the subspace M_μ spanned by the eigenvectors of T associated with the eigenvalue μ.

Then

(i) $E_\lambda P_\mu = \begin{cases} P_\mu, & \lambda \geq \mu \\ 0, & \lambda < \mu \end{cases}$

(ii) for $\epsilon > 0$

$$E_\mu\psi - E_{\mu-\epsilon}\psi \to P_\mu\psi$$

as $\epsilon \to 0$ for any $\psi \in H$,

Proof: For any $\epsilon > \delta > 0$ and any $\psi \in H$ set

$$\pi_\epsilon := E_\mu - E_{\mu-\epsilon}, \quad \pi_\delta := E_\mu - E_{\mu-\delta}$$

then

$$\| \pi_\epsilon\psi - \pi_\delta\psi \|^2 = \| \pi_\epsilon\psi \|^2 - (\psi, \pi_\delta\pi_\epsilon\psi) - (\psi, \pi_\epsilon\pi_\delta\psi) + \| \pi_\delta\psi \|^2 .$$

Since $\epsilon > \delta > 0$ and $\{E_\lambda\}$ is a spectral family we have

$$\pi_\epsilon\pi_\delta = \pi_\delta = \pi_\delta\pi_\epsilon.$$

Consequently

$$\| \pi_\epsilon\psi - \pi_\delta\psi \|^2 = \| \pi_\epsilon\psi \|^2 - \| \pi_\delta\psi \|^2 . \tag{4.34}$$

Since the left-hand side of (4.34) is positive we can conclude that $\| \pi_\epsilon\psi \|^2$ is monotonic decreasing as $\epsilon \to 0$. As $\| \pi_\epsilon\psi \|^2$ is bounded this implies that $\| \pi_\epsilon\psi \|^2$ tends to a limit as $\epsilon \to 0$. Consequently, the Cauchy property of convergent sequences implies

$$\| \pi_\epsilon\psi \|^2 - \| \pi_\delta\psi \|^2 \to 0 \qquad \text{as } \epsilon, \delta \to 0. \tag{4.35}$$

Thus (4.34) implies that the elements $\pi_\epsilon\psi$ also have the Cauchy property. Since H is complete it therefore follows that $\pi_\epsilon\psi$ tends to a limit in H as $\epsilon \to 0$. Denote this limit by

$$\pi_\epsilon\psi \to \psi_\mu \quad \text{as } \epsilon \to 0. \tag{4.36}$$

If $\lambda \geq \mu$ then

$$E_\lambda\pi_\epsilon = E_\lambda(E_\mu - E_{\mu-\epsilon}) = E_\mu - E_{\mu-\epsilon} = \pi_\epsilon.$$

If $\lambda < \mu$ with ϵ small enough to ensure $\lambda < \mu - \epsilon$, then

$$E_\lambda\pi_\epsilon = E_\lambda(E_\mu - E_{\mu-\epsilon}) = E_\lambda - E_\lambda = 0.$$

Therefore

$$E_\lambda\psi_\mu = \begin{cases} \psi_\mu, & \lambda \geq \mu \\ 0, & \lambda < \mu. \end{cases} \tag{4.37}$$

Consequently, on occasionally writing $\psi_\mu \equiv \psi(\mu)$ to emphasise the dependence on the spectral parameter μ and on changing dummy variables under integrals where necessary, we obtain

$$\begin{aligned}(\phi, T\psi_\mu) &= \int_{-\infty}^{\infty} \lambda d_\lambda(\phi, E_\lambda\psi(\mu)) = \int_{-\infty}^{\infty} \lambda d_\lambda \int_{-\infty}^{\infty} \phi(\mu)\overline{E_\lambda\psi(\mu)}d\mu \\ &= \int_{-\infty}^{\infty} \lambda d_\lambda \int_{-\infty}^{\lambda} \phi(\mu)\overline{\psi(\mu)}d\mu = \int_{-\infty}^{\infty} \phi(\lambda)\lambda\overline{\psi(\lambda)}d\lambda \\ &= (\phi, \mu\psi_\mu).\end{aligned}$$

Since this result holds for any $\phi \in H$ it follows that

$$T\psi_\mu = \mu\psi_\mu.$$

Thus, provided $\psi_\mu \in H$ is non-trivial, μ is an eigenvalue of T with associated eigenvector ψ_μ. Furthermore, using (4.29) we obtain, for any $\phi, \psi \in H$,

$$\begin{aligned}(\phi, T^2\psi) &= \int_{-\infty}^{\infty} \lambda d_\lambda(\phi, E_\lambda T\psi) = \int_{-\infty}^{\infty} \lambda d_\lambda(E_\lambda\phi, T\psi) \\ &= \int_{-\infty}^{\infty} \lambda d_\lambda \int_{-\infty}^{\infty} y d_y(E_\lambda\phi, E_y\psi) = \int_{-\infty}^{\infty} \lambda d_\lambda \int_{-\infty}^{\infty} y d_y(\phi, E_\lambda E_y\psi) \quad (4.38) \\ &= \int_{-\infty}^{\infty} \lambda d_\lambda \int_{-\infty}^{\lambda} y d_y(\phi, E_y\psi) \\ &= \int_{-\infty}^{\infty} \lambda^2 d_\lambda(\phi, E_\lambda\psi).\end{aligned}$$

Suppose now that μ is an eigenvalue of T. Then

$$\begin{aligned}\int_{-\infty}^{\infty} (\lambda^2 - 2\mu\lambda + \mu^2)d(\phi, E_\lambda P_\mu\psi) &= (\phi, [T^2 - 2\mu T + \mu^2]P_\mu\psi) \\ &= 0 \qquad (4.39)\end{aligned}$$

for any $\phi, \psi \in H$. This follows from (4.29) and the fact that for any $\psi \in H$ we have $P_\mu\psi = \eta$ where $T\eta = \mu\eta$.

We would remark that in obtaining (4.39) we have used the fact that for $\{E_\lambda\}$, the spectral family of T, there holds

$$\int_{-\infty}^{\infty} d(\phi, E_\lambda\psi) = (\phi, \psi) \quad \forall\phi, \psi \in H. \qquad (4.40)$$

This follows immediately from the properties of Riemann-Stieltjes integrals and the spectral family since

$$\int_{-\infty}^{\infty} d(\phi, E_\lambda\psi) = \lim_{\substack{a\to+\infty \\ b\to-\infty}} [(\phi, E_\lambda\psi)]_b^a = (\phi, \psi).$$

In the particular case when $\phi = P_\mu\psi$ we obtain from (4.39)

$$\int_{-\infty}^{\infty} (\lambda - \mu)^2 d \parallel E_\lambda P_\mu \psi \parallel^2 = 0. \tag{4.41}$$

Now $\parallel E_\lambda P_\mu \psi \parallel^2$ is a monotonic increasing function of λ. To see this simply recall that $\{E_\lambda\}$ is a spectral family and $\lambda_1 \leq \lambda_2$ implies $E_{\lambda_1} \leq E_{\lambda_2}$. Consequently, recalling the definition of a Riemann-Stieltjes integral, (4.20), we infer from (4.41) that the only place where $\parallel E_\lambda P_\mu \psi \parallel^2$ could possibly change value is where $\lambda = \mu$. To see that this is indeed the case we first notice that the properties of the spectral family indicate that

$$\parallel E_\lambda P_\mu \psi \parallel^2 \rightarrow \begin{cases} 0 & \text{as } \lambda \rightarrow -\infty \\ \parallel P_\mu \psi \parallel^2 & \text{as } \lambda \rightarrow +\infty \end{cases}$$

and therefore

$$E_\lambda P_\mu \psi = \begin{cases} P_\mu \psi, & \lambda \geq \mu \\ 0, & \lambda < \mu \end{cases}$$

for any $\psi \in H$. Consequently for any $\psi \in H$

$$(E_\mu - E_{\mu-\epsilon}) P_\mu \psi = (E_\mu P_\mu - E_{\mu-\epsilon} P_\mu)\psi = P_\mu \psi \tag{4.42}$$

which indicates that E_λ does indeed jump in value at $\lambda = \mu$ an eigenvalue.

To complete the proof of the theorem we notice that for any $\psi \in H$

$$\begin{aligned} \parallel \pi_\epsilon (I - P_\mu)\psi \parallel^2 &= (\pi_\epsilon (I - P_\mu)\psi, \pi_\epsilon (I - P_\mu)\psi) \\ &= ((I - P_\mu)\psi, \pi_\epsilon^2 (I - P_\mu)\psi) \\ &= ((I - P_\mu)\psi, \pi_\epsilon (I - P_\mu)\psi) \\ &\rightarrow 0 \quad \text{as} \quad \epsilon \rightarrow 0. \end{aligned} \tag{4.43}$$

This result follows because, as we have seen in (4.36), $\pi_\epsilon (I - P_\mu)\psi$ tends to a limit as $\epsilon \rightarrow 0$. If this limit is not zero then, by definition of π_ϵ, it must be an eigenvector of T which is orthogonal to $(I - P_\mu)\psi$. Therefore, using (4.42) we obtain for any $\psi \in H$,

$$\begin{aligned} (E_\mu - E_{\mu-\epsilon})\psi &= \pi_\epsilon P_\mu \psi + \pi_\epsilon (I - P_\mu)\psi \\ &= P_\mu \psi + \pi_\epsilon (I - P_\mu)\psi \\ &\rightarrow P_\mu \psi \quad \text{as} \quad \epsilon \rightarrow 0. \end{aligned}$$

by (4.43). ∎

A similar result can be obtained for unitary operators.

Theorem 4.20

Let $U : H \to H$ be a linear, unitary operator with an associated spectral family $\{E_\lambda\}$ and spectral decomposition

$$U = \int_0^{2\pi} e^{i\lambda} dE_\lambda.$$

Then E_λ has a discontinuity at $\lambda = \theta$ if and only if $e^{i\theta}$ is an eigenvalue of U.

Let $P_\theta : H \to M_\theta$ denote the projection operator onto the subspace M_θ spanned by the eigenvectors of U associated with the eigenvalue $e^{i\theta}$.

Then

(i) $E_\lambda P_\theta = \begin{cases} P_\theta, & \lambda \geq \theta \\ 0, & \lambda < \theta \end{cases}$

(ii) for $\epsilon > 0$

$$E_\theta \psi - E_{\theta-\epsilon}\psi \to P_\theta \psi$$

as $\epsilon \to 0$ for any $\psi \in H$.

Proof: The proof of this theorem is virtually the same as for the preceding theorem. The main difference is centred on showing that $\| E_\lambda P_\theta \psi \|^2$ can only change its value when $\lambda = \theta$ an eigenvalue. This is established by noticing

$$\begin{aligned} \int_0^{2\pi} | e^{i\lambda} - e^{i\theta} |^2 \, d \| E_\lambda P_\theta \psi \|^2 &= 2\int_0^{2\pi} d \| E_\lambda P_\theta \psi \|^2 - 2Re\{e^{i\theta} \int_0^{2\pi} e^{i\lambda} d \| E_\lambda P_\theta \psi \|^2\} \\ &= 2 \| P_\theta \psi \|^2 - 2Re\{e^{-i\theta} \int_0^{2\pi} e^{i\lambda} d(P_\theta \psi, E_\lambda P_\theta \psi)\} \\ &= 2 \| P_\theta \psi \|^2 - 2Re\{e^{-i\theta}(P_\theta \psi, U P_\theta \psi)\} \\ &= 0. \end{aligned}$$

■

These two Theorems indicate that the jumps in the value of E_λ are the same as in the finite-dimensional case. However, in an infinite-dimensional setting a continuous increase in E_λ is also possible as we have already pointed out in the remark after Example 4.18. We shall clarify this situation below. First we need the following result.

Theorem 4.21

Let $T : H \to H$ be a linear self-adjoint operator. A number λ belongs to the $\rho(T)$, the resolvent set of T if and only if there exists a constant $c > 0$ such that for every $\phi \in H$

$$\| (T - \lambda I)\phi \| \geq c \| \phi \| .$$

Proof: This follows immediately from Definition 4.1 and the proof of Theorem 2.91.∎

We now show that the resolvent set of a self-adjoint operator can also be characterised in terms of properties of the spectral family.

Theorem 4.22

Let T and $\{E_\lambda\}$ be as in Theorem 4.19. A real number μ belongs to $\rho(T)$, the resolvent set of T, if and only if there exists a constant $c > 0$ such that $\{E_\lambda\}$ is constant on the interval $[\mu - c, \mu + c]$.

Proof: Assume $\mu \in \mathbf{R}$ and that $\{E_\lambda\}$ is constant on the interval $\Delta := [\mu - c, \mu + c]$ for some constant $c > 0$. Then by Theorems 4.16 and 4.19, together with (4.38) and (4.40)

$$\begin{aligned} \| (T - \mu I)\phi \|^2 &= ((T - \mu I)^2\phi, \phi) \\ &= ((T^2 - 2\mu T + \mu^2 I)\phi, \phi) \\ &= \int_{-\infty}^{\infty} (\lambda - \mu)^2 d(\phi, E_\lambda \phi). \end{aligned} \tag{4.44}$$

When integrating with $\lambda \in \Delta$ then $\{E_\lambda\}$ is a constant and the contribution to the integral in (4.44) is zero. When integrating with $\lambda \notin \Delta$ then $(\lambda - \mu)^2 \geq c^2$. Hence we obtain

$$\| (T - \mu I)\phi \| \geq c^2 \int_{-\infty}^{\infty} d(\phi, E_\lambda \phi) = c^2(\phi, \phi) = c^2 \| \phi \|^2 .$$

It then follows, by Theorem 4.21,that $\mu \in \rho(T)$.

Conversely assume that $\mu \in \rho(T)$. Then there is a constant $c > 0$ such that

$$\| (T - \mu I)\phi \| \geq c \| \phi \| .$$

Using (4.40) and (4.44) this can be re-written in the form

$$\int_{-\infty}^{\infty} (\lambda - \mu)^2 d(\phi, E_\lambda \phi) \geq c^2 \int_{-\infty}^{\infty} d(\phi, E_\lambda \phi). \tag{4.45}$$

If now we assume that $\{E_\lambda\}$ is not constant on Δ, then we can find a constant $\eta \leq c$ such that

$$E_{\mu+\eta} - E_{\mu-\eta} \neq 0$$

because by the properties of a spectral family

$$E_x \leq E_y \quad \text{for } x < y.$$

Hence there exist $\phi, \psi \in H$ such that

$$\phi = (E_{\mu+\eta} - E_{\mu-\eta})\psi \neq 0$$

and we have

$$E_\lambda \phi = E_\lambda (E_{\mu+\eta} - E_{\mu-\eta})\psi.$$

When $\lambda < \mu - \eta$ this equation reduces to

$$E_\lambda \phi = (E_\lambda - E_\lambda)\psi = 0,$$

whilst for $\lambda > \mu + \eta$ it reduces to

$$E_\lambda \phi = (E_{\mu+\eta} - E_{\mu-\eta})\psi.$$

Consequently, the range of integration in (4.45) reduces to

$$\Delta_1 = [\mu - \eta, \mu + \eta].$$

When $\lambda \in \Delta_1$, straightforward calculation indicates

$$(E_\lambda \phi, \phi) = ((E_\lambda - E_{\mu-\eta})\psi, \psi).$$

Hence (4.45) becomes

$$\int_{\mu-\eta}^{\mu+\eta} (\lambda - \mu)^2 d(\phi, E_\lambda \phi) \geq c^2 \int_{\mu-\eta}^{\mu+\eta} d(\psi, E_\lambda \psi).$$

However this inequality is not possible because

$$(\lambda - \mu)^2 \leq \eta^2 \leq c^2.$$

Thus we have a contradiction and the proof is complete ∎

This Theorem indicates that $\mu \in \sigma(T)$ if and only if $\{E_\lambda\}$ is not constant in any neighbourhood of $\mu \in \mathbf{R}$. We can in fact deduce more but in order to do so we need the following important property of self-adjoint operators.

Theorem 4.23

Let $T : H \to H$ be a bounded, linear self-adjoint operator. Then $\sigma_R(T)$, the residual spectrum of T, is empty.

Proof: We show first that if $T : H \to H$ is a linear, densely defined operator then $\lambda \in \sigma_p(T)$ implies $\bar{\lambda} \in \sigma_R(T^*)$. This follows because if $\lambda \in \sigma_R(T)$ then $R_\lambda(T) = (T - \lambda I)^{-1}$ exists either as a bounded or unbounded operator and $\overline{D(R_\lambda(T))} \neq H$. This implies that there exists a non-trivial $\phi \in H$ such that for all $\psi \in D(T - \lambda I) = D(T)$ we have

$$0 = ((T - \lambda I)\psi, \phi) = (T\psi, \phi) - \lambda(\psi, \phi) = (\psi, T^*\phi) - (\psi, \bar{\lambda}\phi)$$

which implies $T^*\phi = \bar{\lambda}\phi$, that is, $\bar{\lambda} \in \sigma_p(T^*)$.

For the given self-adjoint operator T this result yields

$$\lambda \in \sigma_R(T) \Rightarrow \bar{\lambda} \in \sigma_p(T^*) \Rightarrow \lambda \in \sigma_p(T)$$

which is a contradiction. Therefore $\sigma_R(T) = \emptyset$. ■

Theorem 4.23 indicates that for T self-adjoint

$$\sigma(T) = \sigma_p(T) \cup \sigma_c(T).$$

Consequently, since points in $\sigma_p(T)$ correspond to discontinuities in $\{E_\lambda\}$, the spectral family of T, the following Theorem is immediate.

Theorem 4.24

Let T and $\{E_\lambda\}$ be as in Theorem 4.19. A real number μ belongs to $\sigma_c(T)$, the continuous spectrum of T, if and only if $\{E_\lambda\}$ is continuous at μ and is not constant in any neighbourhood of $\mu \in \mathbf{R}$.

4.7 Functions of an operator

Spectral decompositions are particularly useful when functions of an operator are being considered.

Let $T : H \to H$ be a linear, self-adjoint operator with the spectral decomposition

$$T = \int_{-\infty}^{\infty} \lambda dE_\lambda.$$

Let f denote a complex valued function of the real variable x. The corresponding function of the operator T, denoted $f(T)$, can be defined by

$$(\phi, f(T)\psi) = \int_{-\infty}^{\infty} f(\lambda) d(\phi, E_\lambda \psi) \tag{4.46}$$

for all $\phi, \psi \in H$. The integral is, for continuous f, the familiar Riemann-Stieltjes integral. If f is bounded on $\sigma(T)$ then equation (4.46) does indeed hold for all $\phi, \psi \in H$. However, if f is unbounded on $\sigma(T)$ then the result has to be stated more carefully to accommodate the domains involved.

That (4.46) is a reasonable definition is illustrated by the following examples.

Example 4.25

(i) For $f(x) = x$ we have $f(T) = T$. This follows because from (4.46)

$$\int_{-\infty}^{\infty} \lambda d(\phi, E_\lambda \psi) = (\phi, T\psi).$$

(ii) For $f(x) = 1$ we have $f(T) = I$. This follows from (4.46) and (4.40) because

$$\int_{-\infty}^{\infty} d(\phi, E_\lambda \psi) = (\phi, \psi).$$

(iii) If f, g are continuous complex valued functions of a real variable x and $(fg)(x) = f(x)g(x)$ then for any $\phi, \psi \in H$

$$\begin{aligned}
(\phi, f(T)g(T)\psi) &= \int_{-\infty}^{\infty} f(\lambda) d_\lambda (\phi, E_\lambda g(T)\psi) \\
&= \int_{-\infty}^{\infty} f(\lambda) d_\lambda \int_{-\infty}^{\infty} g(\mu) d_\mu (E_\lambda \phi, E_\mu \psi) \\
&= \int_{-\infty}^{\infty} f(\lambda) d_\lambda \int_{-\infty}^{\lambda} g(\mu) d_\mu (\phi, E_\mu \psi) \\
&= \int_{-\infty}^{\infty} f(\lambda) g(\lambda) d_\lambda (\phi, E_\lambda \psi) = \int_{-\infty}^{\infty} (fg)(\lambda) d_\lambda (\phi, E_\lambda \psi).
\end{aligned}$$

Hence

$$(fg)T = f(T)g(T).$$

(iv) For any element $\phi \in H$

$$(\phi, f(T)\phi) = \int_{-\infty}^{\infty} f(\lambda) d \parallel E_\lambda \phi \parallel^2 .$$

(v) For f such that $f(x)^* = f^*(x)$ then

$$\begin{aligned}\parallel f(T)\psi \parallel^2 &= (\psi, [f(T)]^* f(T)\psi) = (\psi, (f^* f)(T)\psi) \\ &= \int_{-\infty}^{\infty} (f^* f)(\lambda) d(\psi, E_\lambda \psi) \\ &= \int_{-\infty}^{\infty} \mid f(\lambda) \mid^2 d \parallel E_\lambda \psi \parallel^2 .\end{aligned}$$

There is a useful relation between self-adjoint and unitary operators. For T and $\{E_\lambda\}$ as in (4.46) define for each $t \in \mathbf{R}$ and for all $\phi, \psi \in H$

$$(\phi, U_t \psi) = \int_{-\infty}^{\infty} e^{-it\lambda} d(\phi, E_\lambda \psi). \tag{4.47}$$

This relation defines the operator

$$U_t = e^{-itT}.$$

Evidently, $U_0 = I$ and U_t is unitary. Furthermore the properties of exponentials indicate that

$$U_t U_s = U_{t+s}$$

for all $t, s \in \mathbf{R}$.

The converse of these results is contained in the following

Theorem 4.26 (Stone's Theorem)

For each $t \in \mathbf{R}$ let U_t be a unitary operator on H,

(a) If

(i) $(\phi, U_t \psi)$ is a continuous function of t for all $\phi, \psi \in H$

(ii) $U_o = I$ and

(iii) $U_t U_s = U_{t+s}$ for all $s, t \in \mathbf{R}$

then there exists a unique self-adjoint operator $T : H \to H$ such that $U_t = \exp(itT)$ for all t.

(b) An element $\psi \in H$ is in $D(T)$ if and only if

$$\lim_{t \to 0} \left\{ \left(\frac{U_t - I}{it} \right) \psi \right\}$$

exists; in which case the limit is $T\psi$.

(c) If a bounded operator commutes with U_t then it also commutes with T.

4.8 Spectral decompositions of H

A description of a physical system requires, in general, three components; the variables or measurable quantities, the values of the variables at any time and the equations governing the evolution of the variables.

In classical mechanics the measurable quantities, or observables as they are sometimes called, are real or complex variables. If a set of these variables is measured then any real or complex function of these variables is also measured.

In more complicated dynamical systems the variables or observables are often represented by linear operators on a Hilbert space. There are many similarities between the two approaches. For example a linear operator T on a Hilbert space H can always be written in the form

$$\begin{aligned} T &= \left(\frac{T + T^*}{2} \right) + i \left(\frac{T - T^*}{2} \right) \\ &= \mathrm{Re}T + i \,\mathrm{Im}\, T. \end{aligned} \tag{4.48}$$

Since $\mathrm{Re}\, T$ and $\mathrm{Im}\, T$ are clearly self-adjoint it follows that non-self-adjoint operators are the analogues of complex variables and self-adjoint operators the analogues of real variables. However, these parallels need to be treated cautiously. The big difference is that functions of operators are more complicated than functions of real or complex variables. Nevertheless we shall see that there is an intimate connection between observables, self-adjoint operators and spectral families which can be used to provide decompositions into simpler parts of not only a given operator, such at T in (4.48), but also of the underlying Hilbert space, the associated spectrum and related quantities. As an illustration recall that for a self-adjoint linear operator $T : H \to H$, the point spectrum (Definition (4.1)) is

$$\sigma_p(T) = \{\lambda \in \mathbf{R} : \exists n.t\, u \in H\ s.t\, Tu = \lambda u\}. \tag{4.49}$$

We define

$$H_p(T) = \text{linear span of all eigenfunctions of } T. \tag{4.50}$$

As in Theorem 4.10 $H_p(T)$ is in fact a subspace of H which is called the **point subspace** with respect to T. Therefore the Projection Theorem allows us to write

$$H = H_p(T) \oplus H_c(T), \quad H_c(T) = H_p(T)^{\perp} \tag{4.51}$$

where $H_c(T)$ is called the **subspace of continuity** with respect to T.

Let

$$P_p : H \to H_p(T) \tag{4.52}$$

denote the projection onto $H_p(T)$. Then for any $f \in H$

$$f = P_p f + (I - P_p) f = P_p f + P_c f \tag{4.53}$$

where

$$P_c = (I - P_p) : H \to H_c(T) \tag{4.54}$$

is a projection, orthogonal to P_p.

Thus in (4.51), (4.53) we have a decomposition of H which is intimately dependent on T. Furthermore, a decomposition of T is also available. From (4.53) we have

$$Tf = TP_p f + TP_c f =: T_p f + T_c f \tag{4.55}$$

where $T_p(T_c)$ viewed as an operator on $H_p(T)(H_c(T))$ is called the **discontinuous (continuous)** part of T.

Also, using (4.53) together with the spectral family $\{E_\lambda\}$ associated with T we obtain

$$(f, E_\lambda f) = (f, E_\lambda P_p f) + (f, E_\lambda P_c f.) \tag{4.56}$$

This result provides a decomposition of integrals such as (4.29). We shall return to this point later. Before doing so we make two observations. First, the analysis so far has involved only Riemann-Stieltjes integrals. A generalisation involving Lebesgue-Stieltjes integrals will be required to ensure necessary completeness properties. Second, it turns out that the continuous spectrum of an operator can be rather unstable under even quite small perturbations. Therefore, since in essence scattering theory relies on perturbation processes, a potentially better behaved decomposition than (4.51) is required. These two observations can be accommodated by generalising the familiar concept of length on $\mathbf{R}$ to that of the **measure** of a set. It will be convenient to introduce this notion in an abstract setting; we shall then consider specific cases.

Definition 4.28

A class $\mathcal{A}$ of subsets of a set X is called a **σ-algebra** in X if

(i) X and $\emptyset$ belong to $\mathcal{A}$,

(ii) any countable union of elements of $\mathcal{A}$ belongs to $\mathcal{A}$,

(iii) the complement of any element of $\mathcal{A}$ is in $\mathcal{A}$.

Let $\mathcal{F}$ be a class of subsets of X. The intersection of all σ-algebras containing $\mathcal{F}$ is also a σ-algebra and is the smallest σ-algebra containing $\mathcal{F}$. This σ-algebra is said to be generated by $\mathcal{F}$.

Definition 4.29

Let $\mathcal{A}$ be a σ-algebra in a space X. The pair $(X, \mathcal{A})$ is called a **measurable space** and the elements of $\mathcal{A}$ are called **measurable sets**.

Example 4.30

Let $X = \mathbf{R}$ and $\mathcal{B}$ be the σ-algebra generated by the open sets of $\mathbf{R}$. The elements of $\mathcal{B}$ are also called **Borel sets** or $\mathbf{R}$. The σ-algebra $\mathcal{B}$ is generated by the set

$$\{(a, \infty) : a \in \mathbf{R}\}.$$

Definition 4.31

Let X be a space and $f : X \to \mathbf{R}$. The function f is said to be **measurable** if for all $\alpha \in \mathbf{R}$ the set

$$\{x \in X : f(x) > a\}$$

is measurable. That is, the inverse image of every open interval (α, ∞) is an element of a σ-algebra on X.

Definition 4.32

Let $(X, \mathcal{A})$ be a measurable space. A **positive measure** on $(X, \mathcal{A})$ is a mapping

$$\mu : \mathcal{A} \to \bar{\mathbf{R}}_+$$

such that

(i) $\mu(\emptyset) = 0$

(ii) for $\{A_k\}$ a countable collection of disjoint elements of $\mathcal{A}$ (that is $A_m \cap A_n = \emptyset$) then

$$\mu\left\{\bigcup_{k=1}^{\infty} A_k\right\} = \sum_{k=1}^{\infty} \mu(A_k).$$

A satisfactory generalisation of the geometric concept of lengths which is defined for intervals in $\mathbf{R}$ is provided by **Lebesgue measure** which is introduced as follows.

If $A \subset \mathbf{R}$ is any subset of $\mathbf{R}$ then $\bar{m}(A)$, the **outer measure** of A, is defined by

$$\bar{m}(A) = \inf \sum_{j \in J} d(I_j)$$

where the infinum if taken all finite or countable collections of **open** intervals $\{I_j : j \in J\}$ such that $A \subseteq \bigcup_{j \in J} I_j$. Here $d(I_j)$ denotes the **length** of the interval I_j and J is the index set. It is clear that $\bar{m}$ has the properties

(i) $\bar{m}(\emptyset) = 0$

(ii) $\bar{m}(A) \geq 0$

(iii) $A \subseteq B \Rightarrow \bar{m}(A) \leq \bar{m}(B)$

(iv) $\bar{m}(x) = 0, \quad x \in \mathbf{R}$.

This last property follows because for any $x \in \mathbf{R}$

$$x \in I_n = (x - \frac{1}{2n}, \; x + \frac{1}{2n})$$

and for any positive n we have $d(I_n) = \frac{1}{n}$.

The **inner measure** of A, denoted $\underline{m}(A)$, is defined by

$$\underline{m}(A) = \sup\{\bar{m}(G) : G \subset A\}$$

where the supremum is taken over all **closed** subsets G of A. For any $A \subset \mathbf{R}$ we have

$$\underline{m}(A) \leq \bar{m}(A).$$

Definition 4.33

A subset $A \subset \mathbf{R}$ is said to be **Lebesgue measurable** when

$$\underline{m}(A) = \bar{m}(A) =: m(A)$$

and $m(A)$ is called the **Lebesgue measure** of A.

Recall that a function F is said to be continuous from the left (right) at x_0 if

$$f(x_0 - 0) = f(x_0) \quad (f(x_0 + 0) = f(x_0)).$$

We now introduce

Definition 4.34

Let f be a real, nondecreasing continuous function from the right and defined on $\mathbf{R}$. A measure μ, defined on Borel sets of $\mathbf{R}$ with the properties

(i) $$\mu((a,b)) = \mu([a+0, b-0]) = f(b-0) - f(a+0) = f(b-0) - f(a)$$

(ii) $$\mu((a,b]) = \mu([a+0, b+0]) = f(b+0) - f(a+0) = f(b) - f(a)$$

(iii) $$\mu([a,b)) = \mu([a-0, b-0]) = f(b+0) - f(a-0)$$

(iv) $$\mu([a,b]) = \mu([a-0, b+0]) = f(b+0) - f(a-0 = f(b) - f(a-0)$$

is called a **Stieltjes measure** generated by the function f.

Example 4.35

(i) Lebesgue measure is the Stieltjes measure generated by the function $f : x \to x$.

(ii) If f is discontinuous at x then $\mu(\{x\}) \neq 0$. In fact

$$\mu(\{x\}) = \mu([x-0, x+0]) = f(x+0) - f(x-0) = f(x) - f(x-0).$$

(iii) If f is constant on $[a, b)$ then $\mu([a, b)) = 0$.

We would remark that functions which are equal everywhere except on sets of measure zero are said to be **equal almost everywhere** (a.e.). Functions which are equal (a.e.) are said to be **equivalent**.

The Lebesgue-Stieltjes integral of a measurable function g is constructed by measuring, with respect to a Lebesgue-Stieltjes measure, sets of the form $g^{-1}[(a, b)]$. The fact that an inverse image is used reflects the need to exhibit a more sensitive dependence on the properties of the integrand; consequently it is the range rather than the domain of the function that is partitioned when defining the integral.

Let f be a measurable function with respect to a Lebesgue-Stieltjes measure μ and denote the associated Lebesgue-Stieltjes integral of f by

$$\int f d\mu.$$

The set of equivalence classes of functions f such that

$$\int | f | d\mu < \infty$$

is denoted by L_1. The collection L_1 is a normed linear space with norm defined by

$$\| f \|_1 = \int | f | d\mu.$$

In order to emphasise the region of integration and the particular measure involved we will write, for example,

$$L_1 \equiv L_1([a, b], d\mu).$$

It can be shown that $L_1([a,b],d\mu)$ is the completion of $C([a,b])$ with respect to a metric ρ defined by

$$\rho(f,g) := \int_a^b |f-g| \, d\mu$$

where the integral here is the usual Riemann integral.

Example 4.36

If μ is continuously differentiable then

$$\int f d\mu = \int f(\frac{d\mu}{dx})dx$$

where dx is Lebesgue measure.

Definition 4.37

Let μ be a Stieltjes measure defined on Borel sets of $\mathbf{R}$.

(i) The set

$$P = \{x \in \mathbf{R} : \mu(\{x\}) \neq 0\}$$

is called the set of **pure points** of the measure μ.

(ii) μ is called a **continuous measure** if it has no pure points.

(iii) μ is called a pure **point measure** if it is not continuous

For any Borel set $X \in \mathbf{R}$ define

$$\mu_p(X) = \mu(P \cap X), \quad \mu_c(X) = \mu(X) - \mu_p(X).$$

Then μ_p is a measure, μ_c is positive and such that $\mu_c(\{x\}) = 0$ for all $x \in X$. Thus we have an indication of the important result.

Theorem 4.38

A Stieltjes measure μ defined on Borel sets of $\mathbf{R}$ can be uniquely decomposed in the form

$$\mu = \mu_p + \mu_c.$$

A typical example of this decomposition is given by (4.56) with $\mu = (f, E_\lambda f)$. The terms on the right hand side of (4.56) are μ_p and μ_c respectively.

Sets of measure zero play a fundamental role in integration theory and the representation of quantities as integrals. However, as we have seen, in Example 4.35, a set may have measure zero with respect to one measure but not another. The situation can be clarified by means of the following.

Definition 4.39

A measure μ_1 is said to be **absolutely continuous with respect to a measure** μ_2 if any set having μ_2-measure zero also has μ_1-measure zero. This will be denoted $\mu_1 << \mu_2$.

If $\mu_1 << \mu_2$ and $\mu_2 << \mu_1$ then the measures are said to be **equivalent.**

An alternative, and for our purposes rather more convenient, form of this definition is

Definition 4.40

A Stieltjes measure μ is said to be **absolutely continuous** with respect to Lebesgue measure x if there exists a function $f \in L_1(\mathbf{R}, dx)$ locally, that is

$$\int_a^b \mid f(x) \mid dx < \infty$$

for any bounded interval $(a, b) \subset \mathbf{R}$, such that for any measurable $g \in L_1(\mathbf{R}, d\mu)$

$$\int g d\mu = \int g f dx.$$

In this case we write $d\mu = f dx$.

A plausible motivation for this definition is afforded by the observation that we can write

$$\int g d\mu = \int g \frac{d\mu}{dm} dm, \quad f = \frac{d\mu}{dm}.$$

A notion related to the absolute continuity of a measure is that of a singular measure.

Definition 4.41

Let

(i) μ_1, μ_2 be two measures defined on a set X

(ii) $X = X_1 \cup X_2, \quad X_1 \cap X_2 = \emptyset$

if $\mu_1(X_2) = 0$ and $\mu_2(X_1) = 0$ then μ_1 is **singular** with respect to μ_2. Equivalently, we can say μ_2 is singular with respect to μ_1.

Of particular relevance later will be the case when one of μ_1, μ_2 is Lebesgue measure.

We shall say that a measure μ is **concentrated** on a set X_1 if $\mu(X_2) = 0$. Combining this with the above definition we see that μ_1 is singular with respect to μ_2 if μ_1 is concentrated on X_1 and μ_2 concentrated on X_2.

Definition 4.42

A measure μ is said to be **discrete** if it is concentrated on a finite or countable set of points, the discrete points of the measure.

Clearly, a discrete measure is a singular measure since the associated set of points has Lebesgue measure zero.

A fundamental result in the theory of measures is the following.

Theorem 4.43 (Lebesgue Decomposition Theorem)

A Stieltjes measure μ can be uniquely decomposed in the form

$$\mu = \mu_{ac} + \mu_s$$

where

μ_{ac} = absolutely continuous with respect to Lebesgue measure.

μ_s = singular with respect to Lebesgue measure.

With this preparation we return to the matter of obtaining an alternative decomposition of the Hilbert space H than that offered by (4.51). Our aim is to obtain a decomposition of H which reflects not only the influence of the operator T involved but also that of the Lebesgue Decomposition Theorem.

Let H be a Hilbert space endowed with a measure μ; typically $L_2([a,b], d\mu)$. Furthermore let $\{E_\lambda\}$ denote the spectral family associated with a self-adjoint operator $T : H \to H$. We now introduce a **spectral measure**, μ_f, on $\mathbf{R}$ as follows. Set

$$I = (a, b]$$

$$E(I) = E_b - E_a.$$

and define, for any $f \in H$,

$$\mu_f(I) := (f, E(I)f) = \| E(I)f \|^2 \tag{4.57}$$

where we emphasise that $(\cdot, \cdot)$ denotes the inner product on $L_2([a, b], d\mu)$. It is readily seen that μ_f is a Stieltjes measure on Borel sets of $\mathbf{R}$.

We now use the Lebesgue Decomposition Theorem and write

$$\mu_f(I) = \mu_{f,ac}(I) + \mu_{f,s}(I)$$

where $\mu_{f,ac}(I)(\mu_{f,s}(I))$ is an **absolutely continuous (singular)** measure with respect to the Lebesgue measure m.

This motivates

Definition 4.44

A function $f \in H$ is said to be **absolutely continuous (singular)**, on the set I, with respect to the self-adjoint operator $T : H \to H$ if

$$\mu_f(I) = \mu_{f,ac}(I), \quad (\mu_f(I) = \mu_{f,s}(I)).$$

The set of all $f \in H$ which are absolutely continuous (singular) with respect to T is denoted $H_{ac}(T)(H_s(T))$ and is called the **subspace of absolute continuity (subspace of singularity)** of T.

In other words, recalling Definition 4.39, we see that f is absolutely continuous with respect to T if $m(I) = 0 \Rightarrow \mu_f(I) = 0$ that is $m(I) = 0 \Rightarrow E(I)f = 0$.

Similarly, recalling Definition 4.41, f is said to be singular with respect to T if there is an interval I_0 such that $m(I_0) = 0 \Rightarrow (1 - E(I_0))f = 0$.

The following result can be obtained.

Theorem 4.45

The subspaces $H_{ac}(T)$ and $H_s(T)$ are subspaces of H, are orthogonal complements of each other in H and reduce T that is

$$H = H_{ac}(T) \oplus H_s(T).$$

Furthermore, if $P_{ac}(T) : H \to H_{ac}(T)(P_s(T) : H \to H_s(T))$ is a projection then T commutes with $P_{ac}(T)(P_s(T))$.

The proof of this theorem is quite straightforward and is left as an exercise. The result itself leads quite naturally to the following notions.

Definition 4.46

The **absolutely continuous (singular)** spectrum of T, denoted $\sigma_{ac}(T)(\sigma_s(T))$ is the spectrum of $TP_{ac}(T)(TP_s(T))$ regarded as an operator in $H_{ac}(T)(H_s(T))$. That is

$$\sigma_{ac}(T) = \sigma(TP_{ac}(T)), \quad \sigma_s(T) = \sigma(TP_s(T)).$$

Both $\sigma_{ac}(T)$ and $\sigma_s(T)$ are closed subsets of $\mathbf{R}$. However they need not be disjoint.

If we assume $\sigma(T) \subseteq (a, b]$ and that there exists a $\lambda \in (a, b)$ such that there is a non-trivial $f \in H$ satisfying $Tf = \lambda f$ then $\lambda \in \sigma_p(T)$. Furthermore, by (4.42) we have

$$E(\{\lambda\})f := \lim_{\epsilon \to 0} (E_\lambda - E_{\lambda-\epsilon})f = f. \tag{4.58}$$

Therefore, using (4.57) we have

$$\mu_f(\{\lambda\}) = \| f \|^2 \neq 0 \tag{4.59}$$

and notice that

$$m(\{\lambda\}) = 0$$

where m denotes Lebesgue measure.

Furthermore, we have

$$\mu_f(\{\lambda\}) = 0, \quad \lambda \notin I = (a, b]. \tag{4.60}$$

and

$$m(\mathbf{R} \setminus I) \neq 0.$$

Collecting the results (4.59) and (4.60) and recalling the comments immediately following Definition 4.44 and Definition 4.46 we see that

$$f \in H_s(T), \quad \lambda \in \sigma_s(T),$$

the latter following because

$$Tf = TP_s f = \lambda f, \quad f \in H_s(T)$$

which implies $\lambda \in \sigma(TP_s)$ and hence that $\lambda \in \sigma_s(T)$.

Consequently,

$$H_p(T) \subseteq H_s(T), \quad \sigma_p(T) \subseteq \sigma_s(T). \tag{4.61}$$

Taking complements throughout (4.61) yields

$$H_{ac}(T) = H_s^{\perp}(T) \subseteq H_p^{\perp}(T) = H_c(T) \tag{4.62}$$

and thus

$$\sigma_{ac}(T) \subseteq \sigma_c(T).$$

Since $H_c(T)$ and $H_{ac}(T)$ are closed subspaces of H we can write

$$H_c(T) = H_{ac}(T) \oplus H_{sc}(T) \tag{4.63}$$

where

$$H_{sc}(T) = H_{ac}^{\perp}(T).$$

Thus we have arrived at the required further decomposition of H

$$H = H_p(T) \oplus H_c(T) = H_p(T) \oplus H_{ac}(T) \oplus H_{sc}(T). \tag{4.64}$$

Comparing (4.64) with the decomposition in Theorem 4.45 we conclude that

$$H_{sc}(T) = H_c(T) \cap H_s(T). \tag{4.65}$$

If $P_{sc}(T) : H \rightarrow H_{sc}(T)$ is a projection then $\sigma(TP_{sc}(T))$, the spectrum of $TP_{sc}(T)$ regarded as an operator in $H_{sc}(T)$, is called the **singularly continuous** spectrum of T and denoted $\sigma_{sc}(T)$. Thus $\sigma_{sc}(T) = \sigma(TP_{sc}(T))$.

Finally, in this section we introduce two further components of the spectrum of a self-adjoint linear operator $T : H \rightarrow H$.

Definition 4.47

(i) The subset of $\sigma(T)$ consisting of all isolated eigenvalues of finite multiplicity is called the **discrete spectrum** of T and denoted by $\sigma_d(T)$.

(ii) The set $\sigma(T) \setminus \sigma_d(T)$ is called the **essential spectrum** of T and denoted by $\sigma_e(T)$.

We see that $\sigma_e(T)$ comprises $\sigma_c(T)$, eigenvalues of infinite multiplicity and points of accumulation of $\sigma_p(T)$.

It is worth noting at this stage that if P_p, P_c are projections of H onto the subspaces $H_p(T)$ and $H_c(T)$ respectively then, since in general we have

$$H = H_p(T) \oplus H_c(T),$$

we have for any $f \in H$

$$\begin{aligned}(f, E_\lambda f) &= (f, E_\lambda[P_p f + P_c f]) \\ &= (f, E_\lambda P f) + (f, E_\lambda P_c f).\end{aligned}$$

On the right hand side the first term is a purely discrete measure whilst the second term is a purely continuous measure (see Definitions 4.42 and 4.37).

§4.9 Examples

In this section we investigate the properties of two classes of operators which will occur frequently in later chapters; specifically, differentiation and multiplication operators. We shall be particularly interested in how these operators are related and in how their spectral properties can be determined. The worked examples we give below illustrate the techniques which can be used when dealing with more complicated operators.

We have already met the differentiation operator and discussed some of its properties but only as an operator on the space of continuous functions (Example 2.87). We now want to study differentiation on $L_2(a,b)(-\infty \leq a < b \leq \infty)$ and for this we shall require more than continuity of the function.

Definition 4.48

A complex-valued function f is said to be **absolutely continuous** (AC) on $[a,b]$ if there exists an integrable function g on $[a, b]$ and a number $\gamma \in \mathbf{C}$ such that

$$f(x) = \int_a^x g(y)dy + \gamma \quad \text{for all } x \in [a, b].$$

Thus we see that if f is absolutely continuous on $[a, b]$ then it is both continuous on $[a, b]$ and differentiable almost everywhere (a.e.) on $[a, b]$ and its derivative coincides with g a.e. on $[a, b]$.

Example 4.49

Let $H = L_2(a, b)$ $(-\infty < a < b < \infty)$ and define the differentiation operator $T_1 : H \to H$ by

$$T_1 f = if' = i\frac{df}{dx} \quad \text{for all } f \in D(T_1)$$

$$D(T_1) = \{f \in H : f \text{ is AC on } [a, b],\ f' \in H,\ f(a) = f(b) = 0\}.$$

Then

(i) $\overline{D(T_1)} = H$

(ii) T_1 is unbounded and symmetric.

Solution: The functions h defined by

$$h_n(x) = x^n, \quad n \geq 0,\ x \in [a, b]$$

are clearly absolutely continuous. Furthermore, their linear span is dense in $L_2(a, b)$. Therefore, if $\epsilon > 0$ is given then cutting off the function x^n in sufficiently small neighbourhoods of a and b we obtain an element $f \in D(T_1)$ such that $\| x^n - f \| < \epsilon$, and we conclude that $\overline{D(T_1)} = H$.

To show that T_1 is unbounded on H introduce the triangular function f_n defined by

$$f_n(x) = \begin{cases} n(x-a) & , \quad x \in [a, a+\frac{1}{n}] \\ 2 - n(x-a) & , \quad x \in [a+\frac{1}{n}, a+\frac{2}{n}] \\ 0 & , \quad x \in [a+\frac{2}{n}, b] \end{cases}$$

where $n \geq 2/(b-a)$. Then straightforward calculation yields

$$\| f_n \|^2 \leq 2/n, \ \| T_1 f_n \|^2 = \| if_n' \|^2 = 2n, \ \frac{\| T_1 f_n \|}{\| f_n \|} \geq n.$$

Hence

$$\| T_1 \| = \sup\{\frac{\| T_1 f \|}{\| f \|} : f \in D(T_1), \; \| f \| \neq 0\} = \infty.$$

The operator T_1 will be symmetric if we can show that

$$(T_1 f, g) = (f, T_1 g) \quad \text{for all } f, g \in D(T_1).$$

This follows by integration by parts since for all $f, g \in D(T_1)$

$$(if', g) - (f, ig') = 0. \tag{4.66}$$

■

Example 4.50

Let the differentiation operator T_1 be defined as in Example 4.49.

Then

(i) $D(T_1^*) = D_1^*$

$D_1^* = \{f \in H : f \text{ is AC on } [a, b], \; f' \epsilon H\}$

(ii) $T_1^* f = if'$ for all $f \in D(T_1^*)$

(iii) $T_1^{**} = T_1$.

Solution: Notice that (4.66) also holds for $f \in D(T_1)$ and $g \in D_1^*$. This then implies that $D_1^* \subseteq D(T_1^*)$ and that

$$T_1^* g = ig' \quad \text{for } g \in D_1^*.$$

It remains to show that $D(T_1^*) \subseteq D_1^*$. To this end suppose given a function $g \in D(T_1^*)$ and define

$$h(x) = \int_a^x (T_1^* g)(y) dy + \gamma \tag{4.67}$$

where $\gamma \in \mathbf{C}$ is determined to ensure

$$\int_a^b (g(y) + ih(y)) dy = 0. \tag{4.68}$$

Clearly h, defined by (4.67) is AC on $[a, b]$. The requirement (4.68) is always possible since $(g + ih)$ is integrable.

Integration by parts yields, for all $f \in D(T_1)$,

$$\begin{aligned}\int_a^b if'(y)\overline{g(y)}dy &= (T_1 f, g) = (f, T_1^* g)\\ &= \int_a^b f(y)\overline{(T_1^* g)(y)}dy = i\int_a^b if(y)\overline{h(y)}dy\end{aligned}$$

and thus

$$\int_a^b if'(y)[\overline{g(y)+ih(y)}]dy = 0. \tag{4.69}$$

This result holds for all $f \in D(T_1)$ so, in particular, it will hold for a function $f \in D(T_1)$ defined by

$$f(x) = \int_a^x [g(y) + ih(y)]dy. \tag{4.70}$$

That such an f is indeed an element of $D(T_1)$ follows since f is clearly AC on $[a, b]$ and

$$\begin{aligned} f' &= g + ih \in H\\ f(a) &= f(b) = 0\end{aligned}$$

the latter following from (4.68). Therefore combining (4.69) and (4.70) we obtain

$$\int_a^b | g(y) + ih(y) |^2 dy = 0.$$

Therefore

$$g(x) = -ih(x) = -i\int_a^x T_1^* g(y)dy + i\gamma$$

for almost all $x \in [a, b]$. Hence

(i) g is AC on $[a, b]$

(ii) $g' = -iT_1^* g \in H$

which imply $g \in D_1^*$. Thus $D(T_1^*) \subseteq D_1^*$ as required.

From Example 4.49 we have that T_1 is symmetric, that is

$$T_1 \subseteq T_1^*. \tag{4.71}$$

Hence, taking adjoints in (4.71) (Theorem 2.94) we find

$$T_1^{**} \subseteq T_1.$$

For T_1^* and T_1^{**} to be well defined we require that $\overline{D(T_1)} = H$ and $\overline{D(T_1^*)} = H$ respectively. Consequently, by definition of the adjoint we can write, meaningfully,

$$(T_1^* g, h) = (g, Th) \quad \text{for all } g \in D(T_1^*),\ h \in D(T_1)$$

and deduce that $h \in D(T_1^{**})$, that is $D(T_1) \subseteq D(T_1^{**})$. Therefore, $T_1 \subseteq T_1^{**}$ and $T_1^{**}h = Th$ for all $h \in D(T)$. Combining this last result with (4.71) we have

$$T_1 \subseteq T_1^{**} \subseteq T_1^*.$$

We now show that $D(T_1^{**}) \subseteq D(T_1)$ and so obtain the required equality.

Let $f \in D(T_1^{**})$ be given. Then

$$(T_1^{**} f, g) = (f, T^* g) \quad \text{for all } g \in D(T_1^*). \tag{4.72}$$

Also we have

$$T_1^{**} f = if'$$

because as we have seen $T_1^{**} \subseteq T_1^*$ which together with (ii) yields the above result.

Using (4.72) we find by integration by parts

$$\begin{aligned} 0 &= (if', g) - (f, ig') \\ &= i(f(b)g(b) - f(a)\overline{g(a)}) \quad \text{for all } g \in D(T^*). \end{aligned}$$

Since this holds for all $g \in D(T_1^*) = D_1^*$ we find, choosing

$$g(x) = \frac{(x-a)}{b-a} \in D_1^*$$

that $g(a) = 0,\ g(b) = 1$ and hence $f(a) = 0$. Similarly choosing

$$g(x) = \frac{(b-x)}{b-a} \in D_1^*$$

we find $g(b) = 0,\ g(a) = 1$ and $f(b) = 0$.

Consequently $f \in D(T_1)$. Hence $D(T_1) \subseteq D(T_1^{**})$ as required. ∎

A further property of T_1 can be deduced by making use of the following result.

Lemma 4.51

Let H be a Hilbert space and $A : H \to H$ a linear operator such that $\overline{D(A)} = H$. The operator A is closed if and only if $\overline{D(A^*)} = H$ and $A = A^{**}$.

Proof: If $A : H \to H$ is such that $\overline{D(A)} = H$ then A^* is a closed operator (Theorem 2.96).

Since $\overline{D(A^*)} = H$ then A^{**} is well defined and, by Theorem 2.96, is a closed operator. Since $A = A^{**}$ by hypothesis, it follows that A is closed. The converse is left as an exercise. ∎

Corollary 4.52

Let T_1 be defined as in Example 4.49. Then T_1 is a closed operator.

Proof: Follows immediately from the working of Example 4.50 and Lemma 4.51. ∎

We now consider differentiation on $H = L_2(-\infty, \infty)$.

Example 4.53

Let $H = L_2(-\infty, \infty)$ and define the differentiation operator $T_2 : H \to H$ by

$$T_2 f = i f' \quad \text{for all } f \in D(T_2)$$

$$D(T_2) = \{f \in H : f \text{ is AC on every finite interval } [a, b]\ (-\infty < a < b < \infty),\ f' \in H\}.$$

Then

(i) $\overline{D(T_2)} = H$

(ii) T_2 is unbounded

(iii) T_2 is self-adjoint.

Solution: The functions

$$h_n(x) = x^n \exp(-x^2/2), \quad n \geq 0$$

are clearly elements of $D(T_2)$. Furthermore, their linear span is dense in $L_2(-\infty, \infty)$ and it follows that $\overline{D(T_2)} = H$.

That T_2 is unbounded follows from the fact that T_2 represents, in particular, the differentiation operator T_1 in $L_2(a,b)$ where $[a,b]$ is any finite closed subinterval of $(-\infty,\infty)$, and T_1 is unbounded (Example 4.49).

To show that T_2 is self-adjoint we first notice that, as before, we can establish that T_2 is symmetric by using (4.66). It thus remains to show that $D(T_2^*) \subseteq D(T_2)$. With this in mind choose any $g \in D(T_2^*)$ and any finite interval $[a,b]$. The working of Example 4.50 can then be repeated to show that g is AC on $[a,b]$ and that $g'(x) = -iT_2^*g(x)$ almost everywhere on $[a,b]$. Since $[a,b]$ was chosen arbitrarily this implies that
$g'(x) = -iT_2^*g(x)$ for almost all $x \in \mathbf{R}$ and that $g' = -iT_2^*g \in H$. Hence $g \in D(T_2)$ which yields $D(T_2^*) \subseteq D(T_2)$ as required. ∎

Example 4.54

The operator T_2 defined in Example 4.53 is closed.

Solution: Follows immediately from Theorem 2.96. ∎

We now turn to a more complete examination of multiplication operators on Hilbert spaces than that given in Example 4.17.

Example 4.55

Let $H = L_2(a,b)$ $(-\infty \leq a < b \leq \infty)$ and define the multiplication operator $Q : H \to H$ by

$$(Qf)(x) = xf(x), \quad f \in D(Q), \ x \in [a,b]$$

$$D(Q) = \{f \in H : xf \in H\}.$$

Then

(i) Q is bounded and $D(Q) = H$ if and only if (a,b) is finite.

(ii) $\overline{D(Q)} = H$.

(iii) Q is self-adjoint.

Solution: Let (a,b) be any finite interval and

$$\gamma := \max\{|a|, |b|\}.$$

Then for every $f \in H$

$$\int_a^b |(Qf)(x)|^2\, dx \leq \gamma^2 \int_a^b |f(x)|^2\, dx = \gamma^2 \| f \|^2 < \infty.$$

Hence Q is a bounded operator and it necessarily follows that $\overline{D(Q)} = H$.

If (a, b) is an unbounded interval then $D(Q)$ contains, at least, all elements in H which vanish outside finite intervals. This latter set is dense in H and it therefore follows that $D(Q)$ is dense in H.

To show that, in this case, Q is an unbounded operator assume for convenience, with no loss of generality, that $a \leq 0,\ b = +\infty$ and introduce a sequence of characteristic functions f_n defined by

$$f_n(x) = \chi([n, n+1))(x) = \begin{cases} 1, & x \in [n, n+1) \\ 0, & x \notin [n, n+1). \end{cases}$$

Then $\| f_n \| = 1$ and $\| Qf_n \|^2 = \int_n^{n+1} y^2 dy \geq n^2, \quad n \geq 0$.

It follows from this last statement that Q is unbounded.

We now notice that for all $f, g \in D(Q)$

$$(Qf, g) = \int_a^b y f(y) \overline{g(y)} dy = \int_a^b f(y) \overline{y g(y)} dy = (f, Qg).$$

From Definition 2.94 it then follows that $g \in D(Q^*)$ and $Q^* g = Qg$. Hence, $D(Q) \subseteq D(Q^*)$ and $Q \subseteq Q^*$. Thus Q is symmetric.

To establish that Q is self-adjoint suppose we are given $g \in D(Q^*)$. Then for every $f \in D(Q)$ the defining relation $(Qf, g) = (f, Q^* g)$ yields

$$\int_a^b f(y) \{\overline{y g(y)} - Q^* g(y)\} dy = 0. \tag{4.73}$$

For any finite interval $(c, d) \subset (a, b)$ define $f \in D(Q)$ by

$$f(x) = \begin{cases} x g(x) - Q^* g(x), & x \in (c, d) \\ 0, & x \notin (c, d). \end{cases}$$

Substituting in (4.73) we obtain $xg(x) = Q^* g(x)$ for almost all $x \in (c, d)$. Since this holds for all subintervals $(c, d) \subset (a, b)$ we obtain $xg(x) = Q^* g(x)$ for almost all $x \in (a, b)$. Also, since by the definition of adjoints we have

$$(Qf, g) = (f, Q^* g) = (f, g^*)$$

it follows that

$$xg = Q^*g = g^* \in H$$

which implies that $g \in D(Q)$. Hence $D(Q^*) \subseteq D(Q)$. Consequently $D(Q) = D(Q^*)$ and $Q = Q^*$. ■

Example 4.56

The operator Q defined in Example 4.55 is closed

Solution: Follows immediately from Theorem 2.96. ■

When developing scattering theories we shall make use of the fact that the differentiation and multiplication operators are often more closely related than would appear at first sight; indeed they can be unitarily equivalent.

Example 4.57

Let $H = L_2(-\infty, \infty)$. Define the differential operator T_2 as in Example 4.53 and the multiplication operator Q as in Example 4.55. If F denotes the Fourier Plancherel operator on H then

$$T_2 = FQF^{-1},$$

that is T_2 and Q are **equivalent** with respect to F.

Solution: We recall that the Fourier Plancherel operator $F : H \to H$ is defined by

$$\begin{aligned}(Ff)(\xi) &= \frac{1}{\sqrt{2\pi}} \int_{\mathbf{R}} e^{-i\xi y} f(y)dy \quad a.a.\ \xi \in \mathbf{R} \\ (F^{-1}f)(\xi) &= \frac{1}{\sqrt{2\pi}} \int_{\mathbf{R}} e^{i\xi y} f(y)dy \quad a.a.\ \xi \in \mathbf{R}\end{aligned} \tag{4.74}$$

It can be shown that F is a unique, unitary operator on H and furthermore

$$\begin{aligned}(Ff)(\xi) &= \frac{1}{\sqrt{2\pi}} \frac{d}{d\xi} \int_{\mathbf{R}} \left\{ \frac{e^{-i\xi y} - 1}{-iy} \right\} f(y)dy \quad a.a.\ \xi \in \mathbf{R} \\ (F^{-1}f)(\xi) &= \frac{1}{\sqrt{2\pi}} \frac{d}{d\xi} \int_{\mathbf{R}} \left\{ \frac{e^{i\xi y} - 1}{iy} \right\} f(y)dy \quad a.a.\ \xi \in \mathbf{R}.\end{aligned} \tag{4.75}$$

Our intention is to show

(i) T_2 and FQF^{-1} are defined on the same domain.

(ii) For every f in this common domain $T_2 f = FQF^{-1}f$.

First we determine $D(FQF^{-1})$.

The domains of T_2 and Q are known (see Examples 4.53 and 4.55 respectively).

Properties of linear operators indicate that

$$D(FQ) = \{f \in D(Q) : Qf \in D(F)\}.$$

However since $D(F) = H$ this imposes no additional restriction on f. Consequently

$$D(FQ) = D(Q).$$

Again

$$D(FQF^{-1}) = \{f \in D(F^{-1}) = H : F^{-1}f \in D(FQ) = D(Q)\}$$

since F is (1-1) then $F^{-1}f \in D(Q)$ is equivalent to $f \in F(D(Q))$ and we see that $D(FQF^{-1}) = F(D(Q))$.

It remains to show that $D(T_2) = F(D(Q))$ and $T_2 f = (FQF^{-1})f$ for all $f \in D(T_2)$.

We notice first that, by (4.75),

$$F^{-1}T_2 f(\xi) = \frac{1}{\sqrt{2\pi}} \frac{d}{d\xi} \int_{\mathbf{R}} \left\{ \frac{e^{i\xi y} - 1}{ih} \right\} if'(y)dy.$$

Integrating by parts and rearranging terms we obtain

$$\begin{aligned} F^{-1}T_2 f(\xi) &= \frac{\xi}{\sqrt{2\pi}} \frac{d}{d\xi} \int_{\mathbf{R}} \left\{ \frac{e^{i\xi y} - 1}{iy} \right\} f(y)dy + \frac{1}{\sqrt{2\pi}} \int_{\mathbf{R}} \left\{ \frac{e^{i\xi y} - 1}{iy} \right\} f(y)dy \\ &\quad + \frac{1}{\sqrt{2\pi}} \int_{\mathbf{R}} \left\{ \frac{iye^{i\xi y} - iy}{y^2} \right\} f(y)dy \\ &= \xi F^{-1}f(\xi) \quad a.a.\ \xi \in \mathbf{R}. \end{aligned}$$

Since $F^{-1}T_2 f \in H$, by definition of $D(T_2)$ and $D(F)$ it follows that $\xi F^{-1}f \in H$ which implies $F^{-1}f \in D(Q)$ and $F^{-1}T_2 f = QF^{-1}f$ for all $f \in D(T_2)$. That is $T_2 f = FQF^{-1}f$ for all $f \in D(T_2)$.

Consequently

$$T_2 \subseteq FQF^{-1}. \tag{4.76}$$

Next we notice that for all $g, h \in D(FQF^{-1}) = F(D(Q))$ we have $F^{-1}g \in D(Q)$ and $F^{-1}h \in D(Q)$. Consequently, since F is unitary and T_2 is self-adjoint we can now obtain

$$(FQF^{-1}g, h) = (QF^{-1}g, F^{-1}h) = (F^{-1}g, QF^{-1}h) = (g, FQF^{-1}h).$$

Thus FQF^{-1} is symmetric. Using this fact together with (4.76) we find

$$T_2 = T_2^* \supseteq (FQF^{-1})^* \supseteq FQF^{-1} \supseteq T_2$$

which implies $D(T_2) = D(FQF^{-1})$ and $T_2 = FQF^{-1}$. ∎

Since F is a unitary operator then T_2 and Q are said to be **unitarily equivalent** with respect to F.

Being able to establish that a certain differential operator is unitarily equivalent to a multiplication operator proves to be of advantage when discussing the spectrum of these operators since the spectrum of the latter is often more readily determined than that of the former. However in general the spectrum of an operator consists of more than just eigenvalues. Consequently a generalisation of Theorem 4.7 is required.

We have seen that when λ is an eigenvalue of a given linear operator $T : H \to H$ then the resolvent $(T - \lambda I)^{-1}$ is a bounded operator on H. When λ is an eigenvalue of T then the associated resolvent does not exist and T behaves like λI on H. There is also the possibility that for some λ the resolvent might exist but be unbounded. The following result can often ease this difficulty.

Theorem 4.58

Let $T : H \to H$ be a linear operator. The following statements are equivalent.

(a) $\lambda \in \mathbf{C}$ is either an eigenvalue of T or if it is not an eigenvalue then $(T - \lambda I)^{-1}$ exists as an unbounded operator on H.

(b) There exists a sequence of unit vectors $\{f_n\}_{n=1}^{\infty} \subset D(T)$ such that

$$\lim_{n \to \infty} (T - \lambda I) f_n = 0. \tag{4.77}$$

Proof: ((a)$\Rightarrow$ (b)): If λ is an eigenvalue and f an associated eigenvector then we obtain (4.77), trivially, by setting $f_n = f$, $n \geq 1$.

If λ is not an eigenvalue and $(T-\lambda I)^{-1}$ is unbounded then there exists a sequence of unit vectors $g \in D((T-\lambda I)^{-1})$ such that

$$\lim_{n\to\infty} \| (T-\lambda I)^{-1} g_n \| = \infty.$$

If now we define

$$f_n = \frac{(T-\lambda I)^{-1} g_n}{\| (T-\lambda I)^{-1} g_n \|}, \quad n \geq 1$$

then

(i) $f_n \in D(T)$ (by definition of resolvent)

(ii) $\| f_n \| = 1$

(iii) $\lim_{n\to\infty} (T-\lambda I) f_n = 0.$

((b)$\Rightarrow$ (a)): Assume there is a sequence $\{f_n\}_{n=1}^{\infty} \subset D(T)$ as in (b). If λ is not an eigenvalue of T we can define

$$g_n = \frac{(T-\lambda I) f_n}{\| (T-\lambda I) f_n \|}.$$

Thus we obtain a sequence $\{g_n\}_{n=1}^{\infty}$ with the properties

(i) $g_n \in D((T-\lambda I)^{-1}), \quad n \geq 1$

(ii) $\lim_{n\to\infty} \| (T-\lambda I)^{-1} g_n \| = \infty.$

Thus the resolvent is unbounded. ∎

This Theorem motivates the following

Definition 4.59

Let $T : H \to H$ be a linear operator. A complex number λ is called a **generalised** (or **approximate**) eigenvalue of T if there exists a sequence of unit vectors $\{f_n\}_{n=1}^{\infty} \subset D(T)$ such that

$$\lim_{n\to\infty} (T-\lambda I) f_n = 0.$$

We can now obtain the required generalisation of Theorem 4.7 in the form

Theorem 4.60

Let H be a Hilbert space and let $R, S : H \to H$ be linear operators unitarily equivalent with respect to a unitary operator U on H.

(i) $\lambda \in \mathbf{C}$ is an eigenvalue of R if and only if λ is an eigenvalue of S.

(ii) If $\lambda \in \mathbf{C}$ is not an eigenvalue of R then $(R-\lambda I)^{-1}$ and $(S-\lambda I)^{-1}$ are unitarily equivalent with respect to U

(iii) $\sigma(R) = \sigma(S)$

(iv) λ is a generalised eigenvalue of R if and only if it is a generalised eigenvalue of S.

Proof: Assume $R = USU^{-1}$. Since $D(U) = H$ it follows that

$$D(R) = \{f \in D(U^{-1}) : U^{-1}f \in D(US)\}$$

but

$$D(US) = \{f \in D(S) : Sf \in D(U) = H\} = D(S)$$

hence $D(R) = UD(S)$.

If $f \in D(S)$ is an eigenvector of S corresponding to the eigenvalue λ then $Uf \in D(R)$ is nontrivial and

$$R(Uf) = USf = \lambda Uf.$$

Therefore Uf is an eigenvector of R corresponding to the same eigenvalue λ. Reverse the argument to obtain (i).

If λ is not an eigenvalue of R then $(R-\lambda I)^{-1}$ exists and

$$(R-\lambda I)^{-1}(R-\lambda I) = I(R-\lambda I) = I(R) \quad (4.78)$$

$$(R-\lambda I)(R-\lambda I)^{-1} = I((R-\lambda I)^{-1}) \quad (4.79)$$

where here $I(R-\lambda I)$ denotes the restriction to the domain of $(R-\lambda I)$ of the identity operator I on H.

Let $T = U^{-1}(R - \lambda I)^{-1}U$, then

$$D(T) = \{f \in D(U) = H : Uf \in D(U^{-1}(R - \lambda I)^{-1})\}$$
$$D(U^{-1}(R - \lambda I)^{-1}) = \{f \in D((R - \lambda I)^{-1}) : (R - \lambda I)^{-1}f \in D(U^{-1}) = H\}$$
$$= D((R - \lambda I)^{-1}).$$

Consequently $D(T) = U^{-1}D((R - \lambda I)^{-1}$.

Hence

$$T(S - \lambda I) = U^{-1}(R - \lambda I)^{-1}UU^{-1}(R - \lambda I)U = I(U^{-1}RU) = I(S)$$
$$(S - \lambda I)T = U^{-1}(R - \lambda I)UU^{-1}(R - \lambda I)^{-1}U = I(R - \lambda I)^{-1} = I(T).$$

Therefore by definition of inverses

$$(S - \lambda I)^{-1} = T = U^{-1}(R - \lambda I)^{-1}U$$

and we obtain (ii).

Thus we see that the resolvent $(S - \lambda I)^{-1}$ is either unbounded or not defined if and only if $(R - \lambda I)^{-1}$ is either unbounded or undefined. Combining this observation with (i) and the definition of generalised eigenvalues we see that (iii) and (iv) follow. ∎

Example 4.61

Let $H = L_2(-\infty, \infty)$ and define the differentiation operator T_2 and multiplication operator Q as in Example 4.57.

Since T_2 and Q are unitarily equivalent (Example 4.57) it follows that $\sigma(T_2) = \sigma(Q)$.

∎

It is often the case that it is simpler to determine $\sigma(Q)$ than $\sigma(T_2)$ in which case Example 4.61 provides the required information about $\sigma(T_2)$. Before determining $\sigma(Q)$ we need some preparation.

Theorem 4.62

If

(i) $T : H \to H$, linearly

(ii) $\overline{D(T)} = \overline{D(T^*)} = H$

(iii) $T = T^{**}$

(iv) $\lambda \in \mathbf{C}$ is given

then $\{f \in D(T) : (T - \lambda I)f = 0\} = \{(T^* - \bar{\lambda}I)D(T^*)\}^{\perp} = \{R(T^* - \bar{\lambda}I)\}^{\perp}$.

Proof:

$$\begin{aligned}(T - \lambda I)f = 0 &\Rightarrow ((T - \lambda I)f, g) = 0 \quad \forall g \in D(T^*) \\ &\Rightarrow (f, (T^* - \bar{\lambda}I)g) = 0 \quad \forall g \in D(T^*) \\ &\Rightarrow f \in \{(T^* - \bar{\lambda}I)D(T^*)\}^{\perp}.\end{aligned}$$

Conversely, we readily see that $f \in \{(T^* - \bar{\lambda}I)D(T^*)\}^{\perp}$ implies that $(T - \lambda I)f = 0$.

Hence we can conclude that λ is an eigenvalue of T if and only if

$$\overline{(T^* - \bar{\lambda}I)D(T^*)} \neq H$$

since this guarantees that there exists a non-trivial $f \in D(T)$ satisfying $(T - \lambda I)f = 0$.■

Corollary 4.63

Let T be self-adjoint. Then λ is an eigenvalue of T if and only if

$$\overline{(T - \lambda I)D(T)} \neq H.$$

If λ is an eigenvalue of T then the corresponding eigenspace is

$$M_\lambda = [(T - \lambda I)D(T)]^{\perp}.$$

Proof: If λ is an eigenvalue of T then $\lambda \in \mathbf{R}$. By Theorem 4.62 we have $\overline{(T - \lambda I)D(T)} \neq H$ and $M_\lambda = \{(T - \lambda I)D(T)\}^{\perp}$.

If λ is not an eigenvalue of T then $\bar{\lambda}$ is not an eigenvalue of T and it follows from Theorem 4.62 that $\overline{(T - \lambda I)D(T)} = H$. ■

Theorem 4.62 and Corollary 4.63 suggest that for a self-adjoint operator the spectrum can be located on the real axis without investigating the resolvent. Indeed, we have seen that λ is not an eigenvalue when $\overline{(T-\lambda I)D(T)} = H$. We can in fact say rather more.

Theorem 4.64

Let $T : H \to H$ be a self-adjoint, linear operator. A complex number λ is a regular value of T if and only if $(T-\lambda I)D(T) = H$.

Proof: If λ is a regular value of T then, by definition

$$(T-\lambda I)D(T) = D((T-\lambda I)^{-1} = H.$$

Conversely, suppose $(T-\lambda I)D(T) = H$. If $\lambda \notin \mathbf{R}$ then since T is self-adjoint it follows that λ must be a regular value. If $\lambda \in \mathbf{R}$ then Corollary 4.63 indicates that λ is not an eigenvalue. Furthermore, $(T-\lambda I)$ is self-adjoint and defined on all of H. Consequently $(T-\lambda I)^{-1}$ is self-adjoint and bounded (see Theorem 2.96). Therefore λ is regular. ∎

Theorem 4.65

The spectrum of a self-adjoint operator $T : H \to H$ consists entirely of generalised eigenvalues.

Proof: Assume $\lambda \in \sigma(A)$ but λ not an eigenvalue. Then

(i) $(T-\lambda I)D(T)$ is dense in H by Corollary 4.61.

(ii) $(T-\lambda I)^{-1}$ is closed since T is self-adjoint by Theorem 2.96.

Consequently, if the resolvent of T were bounded then its domain $(T-\lambda I)D(T)$ would be closed and therefore coincide with H. This would imply that λ was a regular value for T, contradicting our original assumption that $\lambda \in \sigma(A)$. Therefore $(T-\lambda I)^{-1}$ must be unbounded and hence, by Theorem 4.48, λ must be a generalised eigenvalue. ∎

Example 4.66

Let $H = L_2(a,b)$ $(-\infty \leq a < b \leq \infty)$ and define the multiplication operator Q as in Example 4.57.

(i) Q has no eigenvalues

(ii) $\sigma(Q) = [a, b]$

(iii) $\sigma(Q)$ consists entirely of generalised eigenvalues.

Solution: (i) For any $\lambda \in \mathbf{C}$ suppose we have some $f \in D(Q)$ such that

$$\| (Q - \lambda I)f \|^2 = \int_a^b | y - \lambda |^2 | f(y) |^2 \, dy = 0.$$

Since $| y - \lambda | > 0$ for almost all $y \in \mathbf{R}$ it follows that $f(y) = 0$ for almost all $y \in \mathbf{R}$ and hence $f = \theta$, the zero element. We conclude that Q has no eigenvalues.

(ii) For any $\lambda \in [a, b]$ and $n > 1$ define

$$\chi_n(y) = \begin{cases} 1, & y \in [\lambda - \frac{1}{n}, \lambda + \frac{1}{n}] \cap [a, b] \\ 0, & \text{elsewhere} \end{cases}$$

then $\chi_n \in D(Q)$ and is the characteristic function of the interval $[\lambda - \frac{1}{n}, \lambda + \frac{1}{n}] \cap [a, b]$. Furthermore $\| \chi_n \| > 0$ and $f_n = \chi_n / \| \chi_n \|$ is a unit element in $D(Q)$. Therefore

$$\begin{aligned} \| (Q - \lambda I)f_n \|^2 &= \int_a^b | y - \lambda |^2 | f_n(y) |^2 \, dy \\ &\leq \frac{1}{n^2} \int_a^b | f_n(y) |^2 \, dy = \frac{1}{n^2} \end{aligned}$$

and it follows (Definition 4.64) that λ is a generalised eigenvalue of Q and thus $\lambda \in \sigma(Q) = [a, b]$.

(iii) Let $\lambda \in \mathbf{C} \backslash [a, b]$ and define $\delta = \inf\{| y - \lambda | : y \in [a, b]\}$. Clearly $\delta > 0$. Furthermore, for any $g \in L_2[a, b] = H$ define

$$f = \frac{g}{x - \lambda}. \tag{4.80}$$

Then

$$\int_a^b | f(y) |^2 \, dy = \int_a^b \frac{| g(y) |^2}{| y - \lambda |^2} dy \leq \frac{1}{\delta^2} \| g \|^2 < \infty$$

and we see that $f \in H$. It now follows, using (4.80), that $xf = g + \lambda f \in H$. Hence $f \in D(Q)$.

If now we are given any $g \in H$ then we can always find an $f \in D(Q)$ such that

$$(Q - \lambda I)f = g$$

namely f defined by (4.80). Hence $(Q - \lambda I) : D(Q) \to H$ onto. Also since Q has no eigenvalues $(Q - \lambda I)$ must be (1-1). Therefore the resolvent $(Q - \lambda I)^{-1}$ exists.

Since

$$D((Q - \lambda I)^{-1}) = R(Q - \lambda I) = (Q - \lambda I)D(Q) = H$$

we have, on using (4.80)

$$f = (Q - \lambda I)^{-1} g = \frac{g}{x - \lambda}$$

and

$$\| (Q - \lambda I)^{-1} g \|^2 = \int_a^b \frac{| g(y) |^2}{| y - \lambda |^2} dy \leq \frac{1}{\delta^2} \| g \|^2$$

which implies that $\| (Q - \lambda I)^{-1} \| < 1/\delta^2$ and hence that λ is a regular value. ∎

Example 4.67

Combining Example 4.57, Theorem 4.58 and Example 4.66 we see that T_2 has no eigenvalues, that $\sigma(T_2)$ coincides with the real axis and that $\sigma(T_2)$ consists entirely of generalised eigenvalues. ∎

4.10 More on spectral decompositions associated with an operator

To fix ideas we confine attention to the multiplication operator Q defined by

$$Q : H \to H = L_2((a,b), d\mu), \quad -\infty \leq a < b \leq \infty \tag{4.81}$$

$$(Qf)(x) = xf(x), \quad f \in D(Q)$$

$$D(Q) = \{f \in H : xf \in H\}$$

where μ is a typical Stieltjes measure.

Recalling (4.50) and (4.51), and the arguments leading to them, we have the following decomposition of H relative to Q

$$H = H_p(Q) \oplus H_c(Q) \tag{4.82}$$

However, in Example 4.66 we have seen that

(i) Q has no eigenvalues

(ii) $\sigma(Q) = [a, b]$

(iii) $\sigma(Q)$ consists entirely of generalised eigenvalues.

Consequently

$$\sigma_p(Q) = \phi \tag{4.83}$$

and since $H_p(Q)$ is the span of all eigenfunctions of Q it follows that $H_p(Q) = \{0\}$ and that (4.82) reduces to

$$H = H_c(Q) = H_p^{\perp}(Q). \tag{4.84}$$

We have seen, in Theorem 4.45, that H also has the decomposition

$$H = H_{ac}(Q) \oplus H_s(Q). \tag{4.85}$$

This decomposition can be obtained by introducing a spectral measure, of the form (4.57), which is a mechanism for "separating" the elements of $H = L_2((a,b), d\mu)$ into absolutely continuous and singular elements; we then arrive at (4.85). To see how all this is achieved we first recall Example 4.55 where we showed that Q was self-adjoint on H. Consequently Q has an associated spectral family

$$\{E_\lambda(Q)\} = \{E_\lambda\}, \tag{4.86}$$

the right hand side being used when there is no risk of confusion. It is readily verified that this spectral family has the form:

$$\begin{aligned}
(E_\lambda f)(x) &= \begin{cases} f(x), & a \leq x \leq \lambda \\ 0, & x > \lambda \end{cases} \quad , \; a \leq \lambda \leq b \\
E_\lambda &= 0, \quad \lambda < a \\
E_\lambda &= I, \quad \lambda \geq b.
\end{aligned} \tag{4.87}$$

In particular we notice that

$$\begin{aligned}
\int_{\mathbf{R}} \lambda d(E_\lambda f, g) &= \int_a^b \lambda d \int_0^\lambda f(x)\overline{g(x)} d\mu(x) \\
&= \int_a^b \lambda f(\lambda)\overline{g(\lambda)} d\mu(\lambda) = (Qf, g).
\end{aligned}$$

For the moment we shall assume that μ is Lebesgue measure.

We now introduce a spectral measure associated with Q, denoted μ_f, and defined by

$$\mu_f(I) = (f, E(I)f) \tag{4.88}$$

where

$$I = (a, b], \qquad E(I) := E_b - E_a$$

and $(\cdot,\cdot)$ denotes the inner product on $L_2((a,b), d\mu)$.

By analogy with Definition 4.44 we say, given $f \in H$, that $f \in H_c(Q)$ whenever μ_f, defined in (4.88), is a continuous function of λ.

For any $\lambda \in I$ the discontinuity of $\mu_f(I)$ at λ is given by

$$\Delta := \lim_{\epsilon \to 0+} \{(f, E_\lambda f) - (f, E_{\lambda-\epsilon} f)\} = (f, E_{\{\lambda\}} f) \tag{4.89}$$

where

$$E_{\{\lambda\}} := s-\lim_{\epsilon \to 0+} (E_\lambda - E_{\lambda-\epsilon}).$$

Since $E_{\{\lambda\}}$ is a projection (4.89) can be written

$$\Delta = \| E_{\{\lambda\}} f \|^2 .$$

Therefore $f \in H_c(Q)$ if and only if $\| E_{\{\lambda\}} f \| = 0$ for all λ. This requirement can be achieved in one or other of two ways.

Case 1: $E_{\{\lambda\}} = \Theta$.

We see, from Theorem 4.22, that this is only possible for $\lambda \in \rho(Q)$. Consequently, we shall ignore this possibility.

Case 2: $E_{\{\lambda\}} f = 0$ for all $\lambda \in \mathbf{R}$.

We recall the notation and results of Theorem 4.19 which indicate

$$E_{\{\lambda\}} \psi := \lim_{\epsilon \to 0+} \{E_\lambda \psi - E_{\lambda-\epsilon} \psi\} = P_\lambda \psi \tag{4.90}$$

where $P_\lambda : H \to M_\lambda$ and M_λ is the subspace spanned by the eigenvectors of T associated with the eigenvalue λ. Consequently, if λ is an eigenvalue of T with associated eigenfunction ψ then

$$E_{\{\lambda\}} \psi = P_\lambda \psi = \psi.$$

Therefore in our case the requirement $E_{\{\lambda\}}f = 0$ together with (4.90) indicates that $f \in M_\lambda^\perp$. Furthermore, if $E_{\{\lambda\}}f = 0$ for all λ then f is orthogonal to all eigenfunctions of Q that is

$$f \in H_p^\perp(Q) = H_c(Q).$$

However, we already know that $E_{\{\lambda\}}f = 0$ for all $\lambda \in [a, b]$ since $\sigma_p(Q) = \phi$. Therefore any given $f \in H$ is such that $f \in H_c(Q)$ and we can conclude as in (4.84) that

$$H = H_c(Q).$$

With regard to the spectrum of Q we say that $\lambda \in \sigma_c(Q)$ whenever $\lambda \in \sigma(QP_c)$ where P_c is the projection

$$P_c : H \to H_c(Q).$$

However, (4.84) implies that $P_c = I$. Consequently we have that $\lambda \in \sigma_c(Q)$ whenever $\lambda \in \sigma(QP_c) = \sigma(Q)$. Therefore

$$\sigma_c(Q) = \sigma(Q) = [a, b]. \tag{4.91}$$

A characterisation of $\sigma(Q)$ can also be given in terms of the absolutely continuous and singular spectrum (Definition 4.46) of the operator Q. We recall from Definition 4.44 that if $m(S)$ denotes the Lebesgue measure of a set S then f is said to be absolutely continuous with respect to the self-adjoint operator Q if $m(S) = 0 \Rightarrow \mu_f(S) = 0$.

Now, for all $\lambda \in [a, b]$ we have that $m(\{\lambda\}) = 0$. Furthermore, we have seen that for all $\lambda \in [a, b]$

$$\mu_f(\{\lambda\}) = \| E_{\{\lambda\}}f \|^2 = 0.$$

As before, we already know that $E_{\{\lambda\}}f = 0$ for all $\lambda \in [a, b]$. Therefore we can conclude that any given $f \in H$ is such that $f \in H_{ac}(Q)$, that is

$$H = H_{ac}(Q). \tag{4.92}$$

With regard to the spectrum of Q we arrive at the conclusion that

$$\sigma_c(Q) = \sigma(Q) = [a, b]. \tag{4.93}$$

Finally, recalling either the remarks following Definition 4.44 or Theorem 4.45 we see that $H_s(Q) = \{0\}$ and

$$\sigma_s(Q) = \phi. \tag{4.94}$$

Combining these several results and using Definition 4.47 we obtain

$$\begin{aligned} \sigma(Q) = \sigma_c(Q) = \sigma_{ac}(Q) = \sigma_e(Q) = [a, b] \\ \sigma_p(Q) = \sigma_s(Q) = \phi. \end{aligned} \tag{4.95}$$

These spectral decompositions depend on the measure used in the various calculations. That is, if we change the measure with which we are working then we change the space $H = L_2((a, b), d\mu)$ and consequently also the operator realisation of multiplication (say) on H. Specifically in terms of two Stieltjes measures μ_j, $j = 1, 2$ we have, for $j = 1, 2$

(i) $L_2((a, b), d\mu_j) =: H_j$

(ii) Multiplication has an operator realisation $Q_j : H_j \to H_j$.

Assume that the case $j = 1$ was that which we have already discussed. For the case $j = 2$ we shall assume that μ_2 is a singularly, continuous, monotone function that is

(i) $\mu_2(x_1) \geq \mu_2(x_2), \quad x_1 \geq x_2$

(ii) $\mu_2'(x) = 0$ a.e. with respect to Lebesgue measure.

For the associated multiplication operator Q_2 we obtain, in the same way as for Q_1

$$\sigma(Q_2) = \sigma_c(Q_2) = [a, b], \sigma_p(Q_2) = \emptyset. \tag{4.96}$$

The associated spectral measure, on sets, has the form

$$\mu_f(S) = (f, E(S)f) = \int_{\sigma(Q_2)} f(x)\overline{E(S)f(x)}d\mu_2(x).$$

Here

$$d\mu_2(x) = \left(\frac{\partial \mu_2}{\partial m}\right) dm(x) = \delta(x - \lambda_j)dm(x) \quad j = 1, 2, \cdots$$

where $m(x)$ denotes Lebesgue measure and δ the Dirac delta. We have assumed for simplicity that the λ_j, $j = 1, 2, \cdots$ label the Lebesgue sets of measure zero where $\mu'(x)$ is non-zero. Consequently, recognising the properties of δ and $\{E_\lambda\}$ we have

$$\mu_f(S) = \int_{\sigma(Q_2)} f(x)\overline{E(S)f(x)}\delta(x - \lambda_k)dm(x).$$

Consider the particular case when $S = \{\lambda\}$, $\lambda \in [a, b]$. Then

$$\mu_f(\{\lambda\}) = \int_{\sigma(Q_2)} f(x)\overline{E_{\{\lambda\}}f(x)}\delta(x - \lambda_j)dm(x) = f(\lambda_j)E_{\{\lambda\}}f(\lambda_j) \quad j = 1, 2, \cdots \tag{4.97}$$

Now $E_{\{\lambda\}} = s - \lim_{\epsilon \to o+} (E_\lambda - E_{\lambda-\epsilon})$. Consequently if λ_j, for fixed j, is such that

$$\lambda - \epsilon < \lambda_j < \lambda, \quad \epsilon > 0 \tag{4.98}$$

then the properties of the spectral family (4.87) indicate that

$$E_\lambda f(\lambda_j) = f(\lambda_j), \quad E_{\lambda-\epsilon} f(\lambda_j) = 0.$$

Hence for fixed j the result (4.97) reduces to

$$\mu_f(\{\lambda\}) = | f(\lambda_j |^2 . \tag{4.99}$$

For all the singular points λ_j which lie in the interval $(a, b]$ an inequality of the form (4.98) can be obtained for all $\lambda \in [a, b]$. Therefore, since in general f is not the zero function we conclude that

$$m(\{\lambda\}) = 0 \quad \text{but} \quad \mu_f(\{\lambda\}) = | f(x_j) |^2 \neq 0.$$

Thus $f \notin H_{ac}(Q_2)$. However, the decompositions (4.64)

$$H = H_p(Q_2) \oplus H_{ac}(Q_2) \oplus H_{sc}(Q_2),$$

together with the above results indicate $H = H_{sc}(Q_2)$.

Arguing as in the previous case we find

$$\sigma_{sc}(Q_2) = \sigma(Q_2 P_{sc}) = \sigma(Q_2) = [a, b].$$

Collecting these various results we obtain

$$\sigma(Q_2) = \sigma_c(Q_2) = \sigma_{sc}(Q_2) = \sigma_e(Q_2) = [a,b]$$
$$\sigma_p(Q_2) = \sigma_{ac}(Q_2) = \emptyset. \tag{4.100}$$

This result should be compared with (4.96).

4.11 On the determination of spectral families

We have seen that any spectral family $\{E_\lambda\}$ determines a self-adjoint operator A by means of the relation

$$A = \int_{-\infty}^{\infty} \lambda \, dE_\lambda.$$

Moreover, different spectral families lead to different self-adjoint operators. However, we have not, as yet, given an explicit formula for determining $E(\lambda)$ from A. This is provided by the celebrated Stone's formula which relates the spectral family of A and the resolvent of A.

Theorem 4.68 (Stone's Formula)

For all $a, b \in \mathbf{R}$ and for all $f, g \in H$, where H denotes a Hilbert space, the spectral family $\{E_\lambda\}$ associated with a self-adjoint operator $A : H \to H$ and the resolvent of A are related as follows

$$([E_b - E_a]f, g) = \lim_{\delta \downarrow 0} \lim_{\epsilon \downarrow 0} \frac{1}{2\pi i} \int_{a+\delta}^{b+\delta} ([R(t+i\epsilon) - R(t-i\epsilon)]f, g)dt$$

where $R(t_{\pm} i\epsilon) = [A - (t \pm i\epsilon)]^{-1}$.

This result is particularly useful in those cases where the resolvent of A is more immediately obtainable than the spectral family. We illustrate its use in the following example.

Example 4.69

The transverse vibrations of a string occupying the region $\Omega = \mathbf{R}^+$ are governed typically by an initial boundary value problem of the form

$$\left(\frac{\partial^2}{\partial t^2} + A\right) u(x,t) = 0, \quad (x,t) \in \Omega \times \mathbf{R}^+ \tag{4.101}$$
$$u(x,0) = u_0(x), \quad u_t(x,0) = u_1(x), \; u(0,t) = 0 \tag{4.102}$$

where $A : H \to H = L_2(\Omega)$ is defined by

$$Au = -u_{xx}, \quad u \in D(A)$$
$$D(A) = \{u \in H : u_{xx} \in H \ st \ u(0,t) = 0\}.$$

We have seen that a solution of this initial value problem can be written in the form

$$u(x,t) = (\cos tA^{\frac{1}{2}})u_0(x) + A^{-\frac{1}{2}}(\sin tA^{\frac{1}{2}})u_1(x). \tag{4.103}$$

For such a representation to be of any practical use we must be able to interpret such terms as $(\cos tA^{\frac{1}{2}})u_0(x)$. This can be done by means of the spectral theorem provided the spectral family, $\{E_\lambda\}$, of A can be determined. This we can do by means of Stone's formula.

To use Stone's formula we first compute the resolvents $R(t \pm i\epsilon)$ of A. To this end we consider the boundary value problem

$$(A - \lambda)v(x) = f(x), \quad v(0) = 0. \tag{4.104}$$

This is an ordinary differential which has a solution given by

$$v(x) = (A - \lambda)^{-1}f(x) = \int_\Omega G(x,y)f(y)dy \tag{4.105}$$

where the Green's function is readily found to have the form

$$G(x,y) = \begin{cases} \dfrac{e^{i\sqrt{\lambda}x}\sin\sqrt{\lambda}y}{\sqrt{\lambda}}, & 0 \le y \le x \\ \dfrac{e^{i\sqrt{\lambda}y}\sin\sqrt{\lambda}x}{\sqrt{\lambda}}, & 0 \le x \le y. \end{cases}$$

We now define

$$\lambda_+ := t + i\epsilon = Re^{i\theta}, \ R = \sqrt{t^2 + \epsilon^2}, \ \theta = \tan^{-1}(\epsilon/t)$$
$$\lambda_- := t - i\epsilon = Re^{-i\theta}.$$

Consequently, choosing

$$\sqrt{\lambda_+} = R^{\frac{1}{2}}e^{i\theta/2}, \quad \sqrt{\lambda_-} = R^{\frac{1}{2}}e^{i(\pi-\theta/2)}$$

we see that in the limit as $\epsilon \downarrow 0$

$$\sqrt{\lambda_+} \to +\sqrt{t}, \quad \sqrt{\lambda_-} \to -\sqrt{t}.$$

Therefore

$$R(t+i\epsilon) - R(t-i\epsilon) = (A-\lambda_+)^{-1} - (A-\lambda_-)^{-1}$$

where, by writing (4.105) out in full we obtain

$$(A-\lambda_\pm)f(x) = \frac{\sin\sqrt{\lambda_\pm}x}{\sqrt{\lambda_\pm}}\int_x^\infty f(y)e^{i\sqrt{\lambda_\pm}y}dy + \frac{e^{i\sqrt{\lambda_\pm}x}}{\sqrt{\lambda_\pm}}\int_0^x f(y)\sin\sqrt{\lambda_\pm}y dy.$$

Consequently

$$\lim_{\epsilon\downarrow 0}[R(t+i\epsilon) - R(t-i\epsilon)]f(x) = \frac{2i\sin\sqrt{t}x}{\sqrt{t}}\int_0^\infty f(y)\sin\sqrt{t}y dy.$$

For convenience at this stage assume $f, g \in C_0[a,b]$, $a < b < \infty$. Stone's formula now yields

$$([E_b - E_a]f, g) = \frac{1}{2\pi i}\int_a^b\int_0^\infty \frac{2i\sin\sqrt{t}x}{\sqrt{t}}\int_0^\infty f(y)\sin\sqrt{t}\ y\ dy\overline{g(x)}\ dx\ dt.$$

If we now change variables setting $s = \sqrt{t}$ and introduce

$$\tilde{f}(s) = \left(\frac{2}{\pi}\right)^{\frac{1}{2}}\int_0^\infty f(y)\sin sy\ dy \tag{4.106}$$

then

$$([E_b - E_a]f, g) = \int_{\sqrt{a}}^{\sqrt{b}} \tilde{f}(s)\overline{\tilde{g}(s)}\ ds. \tag{4.107}$$

It can be shown that $\sigma(A) \subset (0,\infty)$. Consequently, $E_a \to 0$ as $a \to 0$. Hence, from (4.107)

$$(E_\lambda f, g) = \int_0^{\sqrt{\lambda}} \tilde{f}(s)\overline{\tilde{g}(s)}ds \tag{4.108}$$

which in turn implies

$$(E_\lambda f)(x) = \left(\frac{2}{\pi}\right)^{\frac{1}{2}} \int_0^{\sqrt{\lambda}} \tilde{f}(s) \sin sx \, ds \tag{4.109}$$
$$=: \int_0^{\sqrt{\lambda}} \tilde{f}(s)\theta_s(x) ds$$

where θ_s has been introduced simply to ease the presentation.

For $f \in D(A)$ the spectral theorem indicates

$$(Af)(x) = \int_0^\infty \lambda dE_\lambda f(x).$$

In the present case we obtain from (4.108)

$$dE_\lambda f(x) = \frac{d}{d\lambda}\left(\int_0^{\sqrt{\lambda}} \tilde{f}(s)\theta_s(x) ds\right) d\lambda$$
$$= \frac{1}{2\sqrt{\lambda}} \tilde{f}(\sqrt{\lambda})\theta_{\sqrt{\lambda}}(x) d\lambda$$
$$= \tilde{f}(s)\theta_s(x) ds, \quad s = \sqrt{\lambda}. \tag{4.110}$$

Hence

$$(Af)(x) = \int_0^\infty \lambda \, dE_\lambda f(x) = \int_0^\infty s^2 \tilde{f}(s)\theta_s(x) ds$$
$$= \left(\frac{2}{\pi}\right)^{\frac{1}{2}} \int_0^\infty s^2 \tilde{f}(s) \sin(sx) ds$$

which we notice is the Fourier sine transform of f.

Furthermore (4.103) now assumes the forms

$$u(x,t) = \cos(tA^{\frac{1}{2}})u_0(x) + A^{-\frac{1}{2}} \sin(tA^{\frac{1}{2}})u_1(x)$$
$$= \int_0^\infty \cos\sqrt{\lambda}\, t \, dE_\lambda u_0(x) + \int_0^\infty \frac{\sin\sqrt{\lambda}t}{\sqrt{\lambda}} dE_\lambda u_1(x)$$
$$= \int_0^\infty \cos st \, \tilde{u}_0(s)\theta_s(x) ds + \int_0^\infty \frac{\sin st}{s} \tilde{u}_1(s)\theta_s(x) ds \qquad (4.111).$$

Once the initial conditions u_0 and u_1 have been given explicitly we can calculate their Fourier sine transforms $\tilde{u}_0$ and $\tilde{u}_1$, respectively, using (4.106), and then obtain from (4.111) a completely determined representation of the associated solution.

An elegant illustration of the full working of (4.111) has been given by Leis who considered the particular case

$$u_0(x) = 0, \quad u_1(x) = 2\left\{\frac{\sin x}{x^2} - \frac{\cos x}{x}\right\}. \tag{4.112}$$

It then follows that

$$\tilde{u}_1(s) = 2s\left(\frac{2}{\pi}\right)^{\frac{1}{2}} \int_0^\infty \frac{\sin x \cos sx}{x} dx =: 2s\left(\frac{2}{\pi}\right)^{\frac{1}{2}} g(s).$$

To compute $g(s)$ we take an arbitrary $\phi \in C_0^\infty(\mathbf{R}^+)$ and consider

$$\begin{aligned}(g, \phi') &= \lim_{R\to\infty} \int_0^R \int_0^\infty \frac{\sin x \cos sx}{x} \phi'(s) ds dx \\ &= \left(\frac{\pi}{2}\right)^{\frac{1}{2}} \int_0^\infty \sin x \tilde{\phi}(x) dx \\ &= \frac{\pi}{2}\phi(1).\end{aligned} \tag{4.113}$$

This last result follows from the fact that the Fourier sine transform is its own inverse. Thus (4.113) is obtained from (4.106) in the particular case $s = 1$. Furthermore

$$(g, \phi') = (-1)(g', \phi) = \frac{\pi}{2}\phi(1) = \frac{\pi}{2}(\phi, \delta(s-1))$$

which implies

$$g'(s) = -\frac{\pi}{2}\delta(s-1). \tag{4.114}$$

Integrating (4.114) we obtain

$$g(s) - g(0) = -\frac{\pi}{2}\int_0^s \delta(\xi - 1) d\xi = -\frac{\pi}{2}H(s-1).$$

Since $g(0) = \dfrac{\pi}{2}$ we can rewrite this in the form

$$g(s) = \frac{\pi}{2}[1 - H(s-1)] = \frac{\pi}{2}H(1-s)$$

and obtain

$$\tilde{u}_1(s) = 2s\left(\frac{\pi}{2}\right)^{\frac{1}{2}} H(1-s).$$

Consequently, in this particular case, (4.111) reduces to

$$
\begin{aligned}
u(x,t) &= 2\int_0^1 \sin(st)\sin sx\,ds = \int_0^1 \{\cos(t-x)s - \cos(t+x)s\}ds \\
&= \frac{\sin(t-x)}{t-x} - \frac{\sin(t+x)}{t+x}
\end{aligned}
$$

which nicely displays the travelling wave components. ∎

Chapter 5
Some applications of semigroup theory

5.1 Introduction and basic results

Our interest when developing scattering theories is centred mainly on initial value problems (IVP) and initial boundary value problems (IBVP) associated with such equations as the wave equation, that is, with equations which are of second order in time. We have seen in Chapter 3 that such equations can be reduced, at least in formal manner, to a system of equations which are of first order in time. It turns out that for these first order equations, results concerning existence, uniqueness and stability can be obtained in an efficient and elegant manner by using the theory of semigroups. To fix ideas let H denote a Hilbert space and consider the IBVP represented by

$$\frac{d\psi(t)}{dt} - G\psi(t) = 0, \quad t \in \mathbf{R}^+, \quad \psi(0) = \psi_0 \tag{5.1}$$

where $\psi \in C^1(\mathbf{R}^+, H)$ and $G : H \supseteq D(G) \to H$. We remark that any boundary conditions (bc) imposed on the problem are incorporated in the definition of $D(G)$ the domain of G.

The first requirement is to clarify the meaning of equation (5.1). The definition of an ordinary derivative indicates that, (5.1) should be interpreted to mean that

(i) $\psi(t) \in D(G), \quad \forall t \in \mathbf{R}^+$

(ii) $\lim_{h \to 0} \| h^{-1}[\psi(t+h) - \psi(t)] - G\psi(t) \| = 0$

where $\| \cdot \|$ denotes the norm in H.

If (5.1) represents a physical, evolutionary system then ideally the IBVP (5.1) should be well-posed in the following sense.

Definition 5.1

An IBVP is well-posed if it has a unique solution which depends continuously on the given data.

This definition implies that small changes in given data should only produce only small changes in the solution.

Let the IBVP (5.1) be well-posed and let $U(t)$ denote a transformation which maps the solution, $\psi(s)$, at time s onto the solution $\psi(s+t)$ at time $s+t$, that is

$$\psi(s+t) = U(t)\psi(s).$$

However, for arbitrary t we have $\psi(t) = U(t)\psi(0) = U(t)\psi_0$. Therefore, since the solution of (5.1) is, by assumption, unique we have

$$U(s+t)\psi_0 = \psi(s+t) = U(t)\psi(s) = U(t)U(s)\psi_0$$

which implies the so-called **semigroup property**

$$U(s+t) = U(t)U(s), \quad s,t \in \mathbf{R}^+. \tag{5.2}$$

Consequently we introduce

Definition 5.2

A family $U := \{U(t) : 0 \leq t < \infty\}$ of bounded, linear operators from a Banach space, $\mathcal{B}$, into itself is called a **strongly continuous, one parameter semigroup**, denoted a C_0**-semigroup**, provided

(i) $\| U(t) \| < \infty$

(ii) $U(t+s)f = U(t)U(s)f = U(s)U(t)f$, for all $f \in \mathcal{B}$, and $t, s \geq 0$

(iii) $U(0)f = f$ for all $f \in \mathcal{B}$

(iv) $t \to U(t)f$; continuous for $t \geq 0$ and all $f \in \mathcal{B}$

If, in addition

(v) $\| U(t)f \| \leq \| f \|$ for all $t \geq 0$ and all $f \in \mathcal{B}$

then U is called a C_0**-contraction** semigroup.

If we recall the interpretation given to the equation (5.1) above then the following definition appears quite natural.

Definition 5.3

The **generator (infinitesimal generator)** G of a C_0-semigroup U is defined by

$$Gf = \lim_{t \to 0+} \frac{U(t)f - f}{t}, \quad f \in D(G) \tag{5.3}$$

if and only if this limit exists.

Example 5.4

The defining properties of a C_0-semigroup suggest, formally at least, that

$$U(t) = \exp(tG), \quad G = \left.\frac{dU(t)}{dt}\right|_{t=0}. \tag{5.4}$$

This also suggests that the solution of (5.1) can be expressed in the form

$$\psi(t) = U(t)\psi_0, \tag{5.5}$$

where U is the semigroup generated by G, provided $\psi_0 \in D(G)$. To see this notice that (5.1) implies

$$G\psi_0 = \left.\frac{d}{dt}(U(t)\psi)\right|_{t=0} + \left.G\psi(t)\right|_{t=0} = G\psi_0.$$

Furthermore we conclude that $U(t)\psi_0$ should be $D(G)$-valued.

To make the statements appearing in this example more precise we first need the following result.

Lemma 5.5

Let $G \in B(H)$. The family

$$U := \{U(t) = \exp(tG) = \sum_{n=0}^{\infty} \frac{(tG)^n}{n!} \;:\; t \in \mathbf{R}^+\}$$

is a C_0-semigroup which satisfies

$$\| U(t) - I \| \to 0 \quad \text{as } t \to 0. \tag{5.6}$$

Moreover, G is the generator of U.

Conversely, if U is a C_0-semigroup satisfying (5.6) then the generator G of U is an element of $B(H)$ and

$$U(t) = \exp tG.$$

Proof: For integers $N, M \geq 0$ and $t \geq 0$ we have

$$\| \sum_{n=N}^{N+M} \frac{(tG)^n}{n!} \| \leq \sum_{n=N}^{N+M} \frac{t^n \| G \|^n}{n!}.$$

Therefore the series $\sum_{n=0}^{\infty} (tG)^n/n!$ converges in the norm of $B(H)$ to an operator $U(t)$. The semigroup property (5.2) follows from the following result for power series

$$\left(\sum_{n=0}^{\infty} \frac{x^n}{n!}\right)\left(\sum_{m=0}^{\infty} \frac{y^m}{m!}\right) = \sum_{k=0}^{\infty} \frac{(x+y)^k}{k!}.$$

It follows immediately from the definition that $U(0) = I$.

Hence

$$\| U(t) - I \| = \| \sum_{n=1}^{\infty} \frac{(tG)^n}{n!} \| \leq \sum_{n=1}^{\infty} \frac{t^n \| G \|^n}{n!} = \exp(t \| G \|) - 1$$

and since the right hand side tends to zero as $t \to 0$ we obtain the **uniform continuity** condition (5.6). Furthermore,

$$\| \frac{U(t) - I}{t} - G \| \leq \| \sum_{n=2}^{\infty} \frac{t^{n-1} G^n}{n!} \| \leq t \| G \|^2 \exp t \| G \|$$

and since the right hand side tends to zero as $t \to 0$ we conclude that G is the generator of U.

A proof of the converse can be found in the two texts referenced in the Commentary. ∎

Corollary 5.6

If $U(t) = \exp(tG)$ then

(i) $\| U(t) \| = \| \exp(tG) \| \leq \exp(t \| G \|), \quad t \in \mathbf{R}^+$.

(ii) $U(t) : \mathbf{R}^+ \to H$ continuously for all $t \in \mathbf{R}^+$.

(iii) $$\frac{d^n}{dt^n}\{U(t)\} = G^n U(t) = U(t)G^n, \quad n = 0, 1, 2, ..., \quad t \in \mathbf{R}^+.$$

The proof of this Corollary is left as an exercise. It should be noticed that the results also hold for $t \in \mathbf{R}$.

In many cases of interest the operator G is unbounded. For such operators the quantities $\| G \|$ and $\exp(tG)$ then become meaningless and results such as Lemma 5.5 have to be modified. With this in mind we first establish

Theorem 5.7

Let $\mathcal{B}$ be a Banach space and $\{U(t) : t \geq 0\} = U \subseteq B(\mathcal{B})$ a C_0-semigroup. Then there exist constants $M \geq 1$ and $\omega \geq 0$ such that

$$\| U(t) \| \leq M \exp(wT), \quad t \in \mathbf{R}^+.$$

Proof Since $U(\cdot)f \in C(\mathbf{R}^+, \mathcal{B})$ it must be bounded with respect to the associated supremum norm. In particular

$$\sup\{\| U(t)f \| : 0 \leq t \leq 1\} < \infty \text{ for all } f \in \mathcal{B},$$

where $\| \cdot \|$ denotes the norm in $\mathcal{B}$. The Uniform Boundedness Principle (Theorem 2.85) implies that there exists a constant M such that

$$\| U(t) \| \leq M, \quad 0 \leq t \leq 1.$$

It follows that $M \geq 1$ because $U(0) = I$.

Now introduce the Gauss symbol $[t]$ to denote the integral part of $t > 0$ and use semigroup properties to obtain

$$U(t) = U([t] + t - [t]) = U([t])U(t - [t]) = U(1)^{[t]}U(t - [t]).$$

If now we fix w by setting $e^w = \| T(1) \|$ then we obtain $\| U(t) \| \leq e^{w[t]}M \leq e^{wt}M$.■

We are now well placed to make Example 5.4 respectable.

Theorem 5.8

Let $\{U(t) : t \geq 0\} = U \subseteq B(\mathcal{B})$ be a C_0-semigroup with generator G. If $f \in D(G)$ then $U(t)f \in D(G)$ for all $t \geq 0$ and

$$\frac{d}{dt}\{U(t)f\} = GU(t)f = U(t)Gf. \tag{5.7}$$

Proof: With $f \in D(G)$ and $s, t, \in \mathbf{R}^+$ together with the commutativity property of semigroups we see that

$$\left\{\frac{U(s) - I}{s}\right\} U(t)f = U(t)\left\{\frac{U(s) - I}{s}\right\} f.$$

Since $U(t)$ is continuous the right-hand side converges, as $s \to 0^+$, with respect to the norm in $\mathcal{B}$ to $U(t)Gf$, a well defined quantity. Thus the left-hand side must also be convergent as $s \to 0^+$. By definition it converges to $GU(t)f$. Combining these results we conclude that $U(t)f \in D(G)$ and

$$GU(t)f = U(t)Gf. \tag{5.8}$$

We must now consider the derivative of $U(t)f$. For $t \geq 0$ the right-hand derivative of $U(t)f$ is given by

$$\lim_{h \to 0+} \left\{\frac{U(t+h) - U(t)}{h}\right\} f = \lim_{h \to 0+} \left\{\frac{U(h) - I}{h}\right\} U(t)f = GU(t)f = U(t)Gf.$$

Similarly, for $t > 0$

$$\begin{aligned}\lim_{h \to 0-} \left\{\frac{U(t+h) - U(t)}{h}\right\} f &= \lim_{s \to 0+} \left\{\frac{U(t) - U(t-s)}{s}\right\} f, \quad h = -s \\ &= \lim_{s \to 0+} \left\{U(t-s)\left[\frac{U(s)f - f}{s} - Gf\right] + U(t-s)Gf\right\}.\end{aligned}$$

In the limit as $s \to 0^+$ the first term vanishes by definition of a generator and the fact that $\| U(t-s) \|$ is bounded on $0 \leq s \leq t$. For the second term notice that

$$\begin{aligned}\| U(t-s)Gf - U(t)Gf \| &= \| U(t-s)Gf - U(t-s)U(s)Gf \| \\ &\leq \| U(t-s) \| \, \| Gf - U(s)Gf \| \\ &\leq M \exp \omega(t-s) \, \| Gf - U(s)Gf \| \\ &\leq M \exp \omega t \, \| Gf - U(s)Gf \| .\end{aligned}$$

The right-hand side of this inequality tends to zero as $s \to 0^+$. Hence $U(t-s)Gf$ tends, strongly, to $U(t)Gf$ as $s \to 0^+$ and we conclude that the left-hand derivative of $U(t)f$ exists and equals $U(t)Gf$. Thus the two sided derivative of $U(t)f$ exists and together with (5.8) yields (5.7). ∎

In the following sections we shall give conditions which ensure that problems of the form (5.1) are well-posed; these we shall see can be expressed in terms of properties of semigroups of operators. First we investigate when the operator G in (5.1) actually generates a semigroup suitable for ensuring that (5.1) is indeed well-posed. We shall then enquire into the consequences of perturbing the operator G with, for example, a potential term.

5.2 On the well-posedness of problems

Whilst much of what we have to say in the following sections can be established in a general Banach space setting we shall confine attention to problems posed in Hilbert spaces. Consequently, let H be a Hilbert space and G a densely defined operator on H. We consider the following IVP

$$\frac{d\psi(t)}{dt} - G\psi(t) = 0, \quad t \in \mathbf{R}^+, \quad \psi(0) = \psi_0 \in D(G). \tag{5.9}$$

A more precise definition of well-posedness than that given in the previous section is as follows.

Definition 5.9

The problem (5.9) is **well-posed** if $\rho(G) \neq \emptyset$ and if for all $\psi_0 \in D(G)$ there exists a unique solution $\psi : \mathbf{R}^+ \to D(G)$ of (5.9) with $\psi \in C^1(\mathbf{R}^+, H)$.

We now establish the following fundamental result.

Theorem 5.10

The problem (5.9) is well-posed if G generates a C_0-semigroup U on H. In this case the solution of (5.9) is given by $\psi(t) = U(t)\psi_0, \quad t \in \mathbf{R}^+$.

Proof: Suppose G generates a C_0-semigroup $U(t)$. If $\psi_0 \in D(G)$ then by Theorem 5.8 we see that $\psi(\cdot) = U(\cdot)\psi_0 \in C^1(\mathbf{R}^+, H)$, is $D(G)$-valued and (5.9) holds. To prove well-posedness it remains to establish uniqueness. To this end let ϕ be any solution of

(5.9). Then, for $0 \leq s \leq t < \infty$ we have

$$\frac{d}{ds}[U(t-s)\phi(s)] = U(t-s)G\phi(s) - U(t-s)G\phi(s) = 0$$

which implies that $U(t-s)\phi(s)$ is independent of s. Consequently, since $U(0) = I$ this independence result allows us to write

$$\phi(t) = U(t-s)\phi(s) = U(t)\phi(0) = U(t)\psi_0.$$

The last equality is obtained because by assumption ϕ satisfies (5.9); in particular the initial condition. Since the right-hand side is in fact $\psi(t)$ we obtain the required uniqueness. ■

The converse of this theorem can also be shown to hold. A proof can be found in the texts cited in the Commentary.

We shall also be interested in non-homogeneous problems of the form

$$\frac{d\psi(t)}{dt} - G\psi(t) = f(t), \quad t \in \mathbf{R}^+, \psi(0) = \psi_0. \tag{5.10}$$

In this connection the following result is available.

Theorem 5.11

Let H be a Hilbert space and $G : H \supseteq D(G) \to H$ be the generator of a C_0-semigroup $U := \{U(t)\}_{t \geq 0} \subseteq B(H)$. If $\psi_0 \in D(G)$ and $f \in C^1(\mathbf{R}^+, H)$ then (5.10) has a unique solution $\psi \in C^1(\mathbf{R}^+, H)$ with values in $D(G)$.

Proof: Let ψ_1, ψ_2 be solutions of (5.10) then $\phi := \psi_1 - \psi_2$ satisfies

$$\phi_t(t) - G\phi(t) = 0, \quad t \in \mathbf{R}^+, \phi(0) = 0. \tag{5.11}$$

Hence by the uniqueness result in Theorem 5.10 we see that $\phi(t) = 0$ for all $t \in \mathbf{R}^+$. Therefore, the problem (5.10) has a unique solution.

We have seen, in Chapter 3, that it is plausible to look for a solution of (5.10) in the form

$$\psi(t) = U(t)\psi_0 + \int_0^t U(t-s)f(s)ds. \tag{5.12}$$

We shall show that (5.12) is indeed a solution of (5.10). To this end set

$$\theta(t) := \int_0^t U(t-s)f(s)ds \equiv \int_0^t U(s)f(t-s)ds. \tag{5.13}$$

The Fundamental Theorem of the Calculus indicates that θ is differentiable on $\mathbf{R}^+$ and

$$\theta'(t) = U(t)f(0) + \int_0^t U(s)f'(t-s)ds. \tag{5.14}$$

Now by assumption U is a C_0-semigroup and f' is continuous. Hence the right-hand side of (5.14) is continuous. Furthermore, by Theorem 5.8, since $\psi_0 \in D(G)$ then $U(t)\psi_0$ is differentiable with

$$\frac{d}{dt}\{U(t)\psi_0\} = U(t)G\psi_0$$

which is clearly continuous in t. Therefore, we see that the right-hand side of (5.12) is continuously differentiable.

We now show that (5.12) implies that $\psi(t) \in D(G)$. Using Theorem 5.8 we see that it will be sufficient to prove that $\theta(t) \in D(G)$. Consequently, by definition of G

$$\begin{aligned}\left\{\frac{U(h)-I}{h}\right\}\theta(t) &= \frac{1}{h}\left\{\int_0^t U(h)U(t-s)f(s)ds - \int_0^t U(t-s)f(s)ds\right\} \\ &\frac{1}{h}\left\{\int_0^t [U(h+t-s)f(s) - U(t-s)]f(s)ds\right\}.\end{aligned} \tag{5.15}$$

The right-hand side denoted by $\mathcal{I}$, can be written

$$\begin{aligned}\mathcal{I} &= \frac{1}{h}\left\{\int_0^{t+h} U(h+t-s)f(s)ds - \int_t^{t+h} U(h+t-s)f(s)ds - \int_0^t U(t-s)f(s)ds\right\} \\ &= \frac{\theta(t+h)-\theta(t)}{h} - \frac{1}{h}\int_t^{t+h} U(h+t-s)f(s)ds.\end{aligned}$$

In the limit as $h \to 0^+$ the first term tends to $\theta'(t)$. The second term we write as

$$J(t,h) = \frac{1}{h}\int_t^{t+h} U(h+t-s)f(s)ds = \frac{1}{h}\int_0^h U(p)f(t+h-p)dp.$$

An application of l'Hôpital's rule then shows

$$\lim_{h\to 0+} J(t,h) = \lim_{h\to 0+}\{U(h)f(t) + \int_0^h U(p)f'(t+h-p)dp\}$$

$$= U(0)f(t) = f(t).$$

Combining these results and using Definition 5.3 we see that $\theta(t) \in D(G)$ and that

$$G\theta(t) = \theta'(t) - f(t) \quad \text{for all } t \in \mathbf{R}^+.$$

Hence we have established that $\psi(t) \in D(G)$ and can obtain

$$\begin{aligned} G\psi(t) &= GU(t)\psi_0 + G\theta(t) \\ &= U(t)G\psi_0 + \theta'(t) - f(t), \quad \text{for all } t \in \mathbf{R}^+. \end{aligned}$$

Furthermore

$$\begin{aligned} \frac{d\psi(t)}{dt} &= \frac{d}{dt}\{U(t)\psi_0 + \theta(t)\} = U(t)G\psi_0 + \theta'(t) \\ &= U(t)G\psi_0 + G\theta(t) + f(t) = G\psi(t) + f(t), \quad t \in \mathbf{R}^+. \end{aligned}$$

Finally, $\psi(0) = U(0)\psi_0 + \theta(0) = \psi_0$ and it follows that (5.12) is a solution of the problem (5.12). ∎

Remark 5.12

We shall see in the next section that if G is the generator of a C_0-semigroup then it is also a closed operator. This will then allow Theorem 5.11 to be established under the condition:

$$f \in C(\mathbf{R}^+, H)$$

with values in $D(G)$ and $Gf \in C(\mathbf{R}^+, H)$ rather than $f \in C^1(\mathbf{R}^+, H)$.

5.3 Generators of semigroups

We have seen that provided we have a C_0-semigroup then we can say something about the well-posedness of an associated IVP. We now want to characterise those linear operators which actually generate C_0-semigroups. In doing this we have to be prepared to work with operators which are not necessarily bounded. This situation is eased by the following result.

Theorem 5.13

Let $U := \{U(t) : t \geq 0\} \subseteq B(H)$ be a C_0-semigroup with generator G. Then $D(G)$ is dense in H. Furthermore $G : H \supseteq D(G) \to H$ is a closed, linear operator.

Proof: Let $f \in H$. We show that for all $t > 0$,

$$g(t) := \int_0^t U(s)f ds \in D(G).$$

To this end let $h > 0$ and consider

$$\left\{\frac{U(h) - I}{h}\right\} \int_0^t U(s)f ds = \frac{1}{h}\int_0^t \{U(s+h)f - U(s)f\}ds. \tag{5.16}$$

The expression on the right can be written

$$\frac{1}{h}\int_h^{t+h} U(p)f dp - \frac{1}{h}\int_0^t U(p)f dp = \frac{1}{h}\int_t^{t+h} U(p)f ds - \frac{1}{h}\int_0^h U(p)f dp.$$

If now we set $w := p - t$ in the first integral then the right-hand side becomes

$$\frac{1}{h}\int_0^h U(w+t)f dw - \frac{1}{h}\int_0^h U(p)f dp.$$

Letting $h \to 0^+$ and using l'Hôpital's Rule this expression then reduces to

$$U(t)f - f. \tag{5.17}$$

Combining (5.16) and (5.17) we obtain

$$Gg(t) = U(t)f - f$$

which clearly indicates that $g(t) \in D(G)$. Now let

$$r(t) = \frac{1}{t}g(t).$$

Then since $D(G)$ is a subspace it follows that we also have $r(t) \in D(G)$. Furthermore, by l'Hôpital's Rule

$$r(t) = \frac{1}{t}\int_0^t U(s)f ds \to U(0)f = f$$

as $t \to 0^+$. Therefore f is the limit of a sequence of elements in $D(G)$. Since f was arbitrary we can conclude that $\overline{D(G)} = H$.

Finally, we want to show that G is a closed operator. Consequently, if $f_n \in D(G)$ is such that

$$f_n \to f \in H \text{ and } Gf_n \to g \in H.$$

then we must show that $f \in D(G)$ and that $Gf = g$.

We notice, by Theorem 5.8, that $\frac{d}{dt}\{U(t)f_n\} = U(t)Gf_n$. This implies that

$$U(t)f_n - f_n = U(t)f_n - U(0)f_n = \int_0^t U(s)Gf_n ds. \tag{5.18}$$

Now since $s \to U(s)$ is continuous on $[0, t]$ then $\| U(s) \|$ is bounded for $0 \leq s \leq t$ and we find that

$$\begin{aligned}\| \int_0^t U(s)Gf_n ds - \int_0^t U(s)g ds \| &= \| \int_0^t U(s)\{Gf_n - g\} ds \| \\ &\leq \int_0^t \| U(s) \| \| Gf_n - g \| ds \leq K \| Gf_n - g \| t\end{aligned}$$

where K is a constant. Since, by assumption, the right-hand side tends to zero as $n \to \infty$ we see that

$$\int_0^t U(s)Gf_n ds \to \int_0^t U(s)g ds \text{ as } n \to \infty.$$

Hence, letting $n \to \infty$ in (5.18) we obtain $U(t)f - f = \int_0^t U(s)g ds$. Therefore $\dfrac{U(t)f - f}{t} = \dfrac{1}{t}\int_0^t U(s)g ds$.

If now we take the limit of this as $t \to 0^+$ then, after one application of l'Hôpital's Rule, we see that the right-hand side tends to g. This implies the existence of the limit, as $t \to 0^+$, of the left hand side. Thus we conclude (see Definition 5.3) that $f \in D(G)$ and that $Gf = g$. Hence G is a closed operator. ∎

With the aid of Theorem we can establish another useful result.

Theorem 5.14

A C_0-semigroup is uniquely determined by its generator.

Proof: Let H be a Hilbert space and

$$U := \{U(t) : t \geq 0\} \subseteq B(H), \quad V := \{V(t) : t \geq 0\} \subseteq B(H)$$

be two C_0-semigroups with the same generator G.

For $t = 0$ the result is trivial since $U(0) = V(0) = I$. Therefore fix $t > 0$. Then for a fixed $f \in D(G)$ define $g : [0,t] \to H$ by

$$g(s) = U(t-s)V(s)f, \quad 0 \leq s \leq t.$$

We notice that $g(0) = U(t)f$ and $g(t) = V(t)f$. Thus we appear to "start" at $U(t)f$ and "finish" at $V(t)f$. However, we claim that we do not really go anywhere; that is we claim that $U(t) = V(t)$. To prove this we show that $g(s)$ is constant on $0 \leq s \leq t$ by examining derivatives of g.

From the semigroup property and Theorem 5.8 we can obtain

$$\frac{d}{ds}g(s) = \frac{d}{ds}\{U(t-s)V(s)f\} = U(t-s)GV(s)f - U(t-s)GV(s)f = 0.$$

Thus $g(s)$ is a constant on $0 \leq s \leq t$ and hence $U(t)f = V(t)f$ for any $f \in D(G)$.

Finally, let $g \in H$. Then by Theorem 5.13 we know that there exists a sequence $\{f_n\} \subset D(G)$ such that $f_n \to f$ as $n \to \infty$. Therefore, by the previous case

$$U(t)f_n = V(t)f_n, \quad n = 1, 2, \cdots.$$

Letting $n \to \infty$ and using the continuity of U and V we obtain $U(t)f = V(t)f$ for all $f \in H$. Thus $U(t) = V(t)$ and the required uniqueness is obtained. ∎

The main result of this section is the following celebrated Theorem.

Theorem 5.15 (Hille-Yosida generation theorem)

A linear operator G in a Hilbert space H generates a C_0-semigroup of **contractions** if and only if

(i) $D(G)$ is dense in H.

(ii) $G : H \supseteq D(G) \to H$ is a closed operator.

(iii) $\rho(G)$, the resolvent set of G, contains all real $\lambda > 0$.

(iv) $R(\lambda, G) := (\lambda I - G)^{-1}$ satisfies

$$\lambda \parallel R(\lambda, G) \parallel \leq 1 \text{ for all } \lambda > 0.$$

Proof: (necessity) Items (i) and (ii) have been dealt with in Theorem 5.14. In order to establish (iii) and (iv) we define an operator $R(\lambda)$ on H according to

$$R(\lambda)f = \int_0^\infty e^{-\lambda t} U(t) f dt, \quad \lambda > 0, \ f \in H.$$

We shall show that $R(\lambda) = (\lambda I - G)^{-1} = R(\lambda, G)$. This might have been expected because if G were a real number than

$$\frac{1}{\lambda - G} = \int_0^\infty e^{-\lambda t} e^{tG} dt.$$

For $f \in D(G)$ and $h > 0$ we have

$$\begin{aligned}
\left\{ \frac{U(t) - I}{h} \right\} R(\lambda)f &= \frac{1}{h} \int_0^\infty e^{-\lambda t} U(t+h) f dt - \frac{1}{h} \int_0^h e^{-\lambda t} U(t) f dt \\
&= \frac{1}{h} \int_h^\infty e^{-\lambda(p-h)} U(p) f dp - \frac{1}{h} \int_0^h e^{-\lambda t} U(t) f dt \\
&\frac{e^{\lambda h}}{h} \left\{ \int_0^\infty - \int_0^h \right\} e^{-\lambda p} U(p) f dp - \frac{1}{h} \int_0^\infty e^{-\lambda t} U(t) f dt \\
&= \frac{(e^{\lambda h} - 1)}{h} \int_0^\infty e^{-\lambda p} U(p) f dp - \frac{e^{\lambda h}}{h} \int_0^h e^{-\lambda p} U(p) f dp
\end{aligned}$$

Taking the limit as $h \to 0^+$ and using l'Hôpital's Rule together with the Fundamental Theorem of the calculus we see that the right-hand side reduces to

$$\lambda \int_0^\infty e^{-\lambda p} U(p) f dp - \lim_{h \to 0+} e^{\lambda h} e^{-\lambda h} U(h) f = \lambda R(\lambda) f - f.$$

Consequently, the limit of the left-hand side exists and on recalling Definition 5.3 we see that $R(\lambda)f \in D(G)$ and $GR(\lambda)f = \lambda R(\lambda) - f$, that is

$$(\lambda I - G) R(\lambda) f = f. \tag{5.19}$$

If we start by assuming that $f \in D(G)$ then

$$R(\lambda)Gf = \int_0^\infty e^{-\lambda t}U(t)Gf dt = \int_0^\infty e^{-\lambda t}GU(t)f dt = GR(\lambda)f,$$

the last line following from the fact that G is the generator of a C_0semigroup (see Theorem 5.8). Hence $R(\lambda)(\lambda I - G)f = (\lambda I - G)R(\lambda)f = f$ and we conclude that

$$R(\lambda)(\lambda I - G)f = f \quad \text{for all } f \in D(G). \tag{5.20}$$

Combining (5.19) and (5.20) we obtain $R(\lambda) = R(\lambda, G)$ for all $\lambda > 0$ as required.

Finally

$$\| R(\lambda, G)f \| = \| R(\lambda)f \| \leq \int_0^\infty e^{-\lambda t} \| U(t) \| \| f \| dt \leq \| f \| \int_0^\infty d^{-\lambda t} dt = \frac{1}{\lambda} \| f \|$$

where we have used the fact that U is a family of contractions.

The proof of suficiency is a straightforward but rather lengthy process. It is found well presented in the texts referred to in the Commentary. ■

Finally in this section we give Stone's Theorem which will be particularly useful for us when developing scattering theories. Before doing so we need a little preparation.

Definition 5.16

A C_0-group on a Hilbert space H is a family of operators $U := \{U(t) : t \in \mathbf{R}\} \subset B(H)$ satisfying the conditions of Definition 5.2 but with $s, t \in \mathbf{R}$. The **generator** of a C_0-group U on H is defined by

$$Gf = \lim_{t \to 0} t^{-1}\{U(t)f - f\}$$

where $D(G)$, the domain of G is the set of all $f \in H$ for which the above limit exists.

We would remark that G is the generator of a C_0-group U if and only if $\pm G$ generate C_0-semigroups $U_\pm$. In which case

$$U(t) = \begin{cases} U_+(t), & t \geq 0 \\ U_-(-t), & t \leq 0. \end{cases} \tag{5.21}$$

We also need a little more information about the properties of adjoint operators in a semigroup theory setting.

Definition 5.17

Let H be a Hilbert space.

(i) An operator $B : H \supseteq D(B) \to H$ is called symmetric if it is densely defined on H and if

$$(Bf, g) = (f, Bg) \text{ for all } f, g \in D(B).$$

(ii) The operator B is **skew-symmetric** if $B \subset -B^*$.

(iii) The operator B is self-adjoint if $B = B^*$.

(iv) The operator B is skew-adjoint if $B = -B^*$.

(v) When H is complex B is skew-adjoint if and only if (iB) is self-adjoint.

The following results can be established and are left as an exercise.

Theorem 5.18

Let H be a Hilbert space and let $G : H \supseteq D(G) \to H$ generate a C_0-semigroup $U := \{U(t) : t \geq 0\} \subset B(H)$. Then $U^* = \{U(t)^* : t \geq 0\}$ is a C_0-semigroup with generator G^*.

The semigroup U is a **self-adjoint** C_0**-semigroup**, that is $U(t)$ is self-adjoint for all $t \in \mathbf{R}^+$, if and only if its generator G is self-adjoint.

Definition 5.19

A C_0-**unitary group** is a C_0-group of unitary operators.

We can now introduce

Theorem 5.20 (Stone's Theorem)

Let H be a Hilbert space. An operator $G : H \supseteq D(G) \to H$ is the generator of a C_0-unitary group U on H if and only if G is skew-adjoint.

5.4 Perturbation of semigroups

We have seen that when developing a scattering theory we tend to regard a given problem (operator) as a perturbation of a somewhat easier problem (operator). In this section we illustrate how these perturbations influence the semigroups associated with these problems. To do this we need the concept of an accretive operator.

Definition 5.21

An operator $G : H \to H$ is called **accretive** if

$$\text{Re}(Gu, u) \geq 0 \text{ for all } u \in D(G).$$

If G is accretive then $(-G)$ is said to be **dissipative.**

The first perturbation result we have is the following.

Theorem 5.22

Let H be a Hilbert space and $G : H \supseteq D(G) \to H$ be the generator of C_0-semigroup. Let B be a dissipative operator on H with $D(B) \supset D(G)$. Assume that there exist constants $0 \leq a < 1,\ b \leq 0$ such that

$$\| Bf \| \leq a \| Gf \| + b \| f \| \quad f \in D(G). \tag{5.22}$$

Then the operator $(G + B)$ with domain $D(G)$ generates a C_0-semigroup.

We shall not prove this Theorem here. It is a standard result in semigroup theory and its proof can be found in the texts cited in the Commentary. An immediate consequence of this Theorem is the following result which will be of interest in later sections.

Theorem 5.23

Let H be a Hilbert space and $G : H \supseteq D(G) \to H$ the generator of a C_0-semigroup. If B is a bounded, linear operator on H then the operator $(G + B)$ also generates a C_0-semigroup.

Chapter 6
More about wave operators

6.1 Introduction

Scattering theory deals with the evolution of states of a system and, in particular, attempts to relate the state of a system as $t \to +\infty$ with the state of that system as $t \to -\infty$. In Chapter 3 we indicated that such a comparison could be made in terms of so-called wave operators. In this chapter we shall introduce these quantities rather more properly and outline some of their more important properties. This we can do most conveniently in an abstract setting; we shall consider specific cases later.

6.2 Abstract evolutionary systems

Let H be a Hilbert space and A, A_0 two linear, self-adjoint operators on H. The operators A and A_0 govern the evolution of the systems of interest and, through their domains of definition, characterise any perturbation involved. We shall consider and compare two systems; a **free system** the evolution of which is described by the IVP

$$\frac{\partial u(x,t)}{\partial t} = -iA_0 u(x,t), \quad u(x,0) = u_0(x) \tag{6.1}$$

and a **perturbed system** which evolves according to

$$\frac{\partial v(x,t)}{\partial t} = -iAv(x,t), \quad v(x,0) = v_0(x). \tag{6.2}$$

In (6.1), (6.2) we assume, for the moment, that $(x,t) \in \mathbf{R}^n \times \mathbf{R}$.The quantities u, v describe the state of the two systems whilst u_0 and v_0 prescribe the initial states of the systems. We have seen, in Chapter 5, that the IVP (6.1) and (6.2) have solutions which can be written in the form

$$u(x,t) = \exp(-itA_0)u_0(x) =: U(t)u_0(x) \tag{6.3}$$

$$v(x,t) = \exp(-itA)v_0(x) =: V(t)v_0(x) \tag{6.4}$$

where $U := \{U(t) : t \in \mathbf{R}\}$ and $V := \{V(t) : t \in \mathbf{R}\}$ are C_0-groups of bounded, linear operators on H. We are interested, when developing a scattering theory, in the behaviour and comparison of the solutions (6.3) and (6.4) as $t \to \pm\infty$.

One possibility of course is that nothing might happen as $t \to \pm\infty$. For instance, consider the case when v_0 is an eigenfunction of A associated with the eigenvalue μ. In this case $Av_0(x) = \mu v_0(x)$ and

$$v(x,t) = V(t)v_0(x) = \exp(-itA)v_0(x) = \exp(-it\mu)v_0(x).$$

Thus we see that $v(x,t)$ does not "evolve away" from the initial state $v_0(x)$ in any way other than by simply a change in phase. This situation should be compared with solutions of the wave equation obtained by separation of variables; such solutions are called **non-propagating** or **stationary solutions** since their nodes are fixed in time. With this analogy in mind the eigenfunctions of A are called the **bound states** or **eigenstates** of the system.

Although bound states appear to be uninteresting as far as scattering theory might be concerned their existence cannot be entirely ignored. However, since, as we have seen in Chapter 4, we have available a decomposition of the form

$$H = H_p(A) \oplus H_c(A)$$

then it is clear that scattering processes are entirely associated with the subspace $H_c(A)$ and the continuous spectrum $\sigma_c(A)$.

Recalling the discussion in Chapter 3 we say that the solutions of (6.1) and (6.2) are asymptotically equal (indistinguishable) if

$$\lim_{t\to\pm\infty} \| v(.,t) - u(.,t) \| = \lim_{t\to\pm\infty} \| V(t)v_0 - U(t)u_0 \| = 0.$$

The self-adjointness of A and A_0 ensures, by means of Stones Theorem (Theorem 5.20), that $U(t)$ and $V(t)$ are unitary operators. Hence

$$\lim_{t\to\pm\infty} \| v(.,t) - u(.,t) \| = \lim_{t\to\pm\infty} \| v_0 - \Omega(t)u_0 \| = \| v_0 - \Omega_{\pm}u_0 \| = 0 \tag{6.5}$$

where

$$\Omega_{\pm} = \lim_{t\to\pm\infty} \Omega(t) = \lim_{t\to\pm\infty} V^*(t)U(t) \tag{6.6}$$

and the limits are understood to be strong limits. We would remark that this definition of the **wave operators**, $\Omega_{\pm}$, is the inverse of that introduced in Chapter 3; see the comment following (3.43). Furthermore, we are assuming here that $\Omega_{\pm}$ are defined on all of H. (See Section 6.5).

We see, from (6.5) that any element v_0 satisfying

$$v_0 = \Omega_+ u_0 \in R(\Omega_+) \tag{6.7}$$

will ensure that

$$\lim_{t \to +\infty} \| v_0 - \Omega_+ u_0 \| = 0. \tag{6.8}$$

An analogous observation can be made involving Ω_-. These remarks introduce the so-called completeness property of wave operators; we shall return to this point in Section 6.5.

Elements v_0, with the property (6.7), define initial states which ensure that the perturbed system (6.2) becomes asymptotically free as $t \to +\infty$. These (initial) states are called **scattering states** for the limit $t \to +\infty$ and $R(\Omega_+)$, the range of Ω_+, is the associated **scattering subspace**. Similar concepts can be introduced for the wave operator Ω_-.

6.3 The scattering operator

The initial state v_0 may be expressed as the sum

$$v_0 = v_0^+ + v_0^\perp \tag{6.9}$$

where $v_0^+ \in R(\Omega_+)$ and $v_0^\perp \in R(\Omega_+)^\perp$. To clarify this situation we first construct projections onto the scattering supspaces introduced in the last section. We notice first that each wave operator is an isometry. To see this let $f, g \in H$. Then we can obtain

$$(\Omega_+ f, \Omega_+ g) = \lim_{t \to +\infty} (\Omega(t) f, \Omega(t) g) = (f, g).$$

This follows because $\Omega(t) = V^*(t) U(t)$ is unitary for fixed t. Hence Ω_+ is an isometry. Similarly, we can establish that Ω_- is an isometry.

Furthermore, since $(\Omega_+ f, \Omega_+ g) = (f, \Omega_+^* \Omega_+ g)$ we can obtain

$$\Omega_+^* \Omega_+ = \Omega_-^* \Omega_- = I. \tag{6.10}$$

If now $g = \Omega_+ f \in R(\Omega_+)$ then $\Omega_+\Omega_+^* g = \Omega_+(\Omega_+^*\Omega_+)f = g$.

Similarly, if $h \in R(\Omega_+)^\perp$ then

$$0 = (h, \Omega_+ f) = (\Omega_+^* h, f) \quad \text{for all } f \in H. \tag{6.11}$$

Consequently, we have $\Omega_+^* h = 0$. Hence $\Omega_+\Omega_+^* h = \Omega_+(\Omega_+^* h) = 0$.

Therefore, we have shown that $\Omega_+\Omega_+^*$ is a projection onto $R(\Omega_+)$. Similarly, we can show that $\Omega_-\Omega_-^*$ is a projection onto $R(\Omega_-)$.

Now consider an initial state v_0 of the perturbed problem which is asymptotically free as $t \to -\infty$, that is $v_0 = \Omega_- u_0$ for a given initial state, u_0, of the free problem. This initial state may be represented in the form

$$v_0 = v_0^+ + v_0^\perp \text{ where } v_0^+ \in R(\Omega_+),\ v_0^\perp \in R(\Omega_+)^\perp. \tag{6.12}$$

Consequently, from the properties of $\Omega_\pm$ developed above

$$v_0^+ = \Omega_+\Omega_+^*(\Omega_- u_0) =: \Omega_+ S u_0 \tag{6.13}$$

and $v_0^\perp \in R(\Omega_+)^\perp$ satisfies $\Omega_+^* v_0^\perp = 0$. Here we have introduced the **scattering operator** S defined by

$$S = \Omega_+^*\Omega_-. \tag{6.14}$$

This operator enables us to relate states which are asymptotically free as $t \to -\infty$ and as $t \to +\infty$. To see this notice that since v_0 is asymptotically free as $t \to -\infty$ then we have

$$\lim_{t \to -\infty} \| V(t)v_0 - U(t)u_0 \| = \| v_0 - \Omega_- u_0 \| = 0.$$

Furthermore, as $t \to +\infty$ we see, by using (6.13) that

$$\lim_{t \to +\infty} \| V(t)v_0^+ - U(t)Su_0 \| = \| v_0^+ - \Omega_+ S u_0 \| = 0. \tag{6.15}$$

This indicates that the state $V(t)v_0$, which is asymptotically free as $t \to -\infty$, has a component, namely $V(t)v_0^+$, which is asymptotically free as $t \to +\infty$.

An alternative way of interpreting S is as follows. Let the perturbed problem have initial data v_0 and assume that the free problem is given data w_+ and w_- which ensure that the perturbed system has a state $v(t) = V(t)v_0$ which is asymptotically free as $t \to +\infty$ and $t \to -\infty$, respectively. It then follows from the above reasoning that

$$\Omega_- w_- = v_0 = \Omega_+ w_+.$$

Consequently, we obtain

$$w_+ = \Omega_+^* \Omega_- w_- =: S w_- \tag{6.16}$$

which relates the initial data required for the free problem which ensure that the perturbed problem is asymptotically free as $t \to +\infty$ and as $t \to -\infty$.

A further connection between S and $\Omega_\pm$ is given by the following result.

Theorem 6.1

(i) $R(\Omega_-) \subseteq R(\Omega_+)$ iff S isometric

(ii) $R(\Omega_+) \subseteq R(\Omega_-)$ iff S^* isometric

(iii) $R(\Omega_+) = R(\Omega)$ iff S unitary.

Proof: Suppose $R(\Omega_-) \subseteq R(\Omega_+)$. For any $f \in H$ we have $\Omega_- f \in R(\Omega_+)$. Now, $\Omega_+ \Omega_+^*$ is a projection onto $R(\Omega_+)$ which implies $\Omega_+ \Omega_+^* (\Omega_- f) = \Omega_- f$. Consequently, since the wave operators are i*sometries we obtain

$$\| Sf \| = \| \Omega_+ Sf \| = \| \Omega_+ \Omega_+^* \Omega_- f \| = \| f \| .$$

Therefore S is isometric.

Conversely, assume that S is isometric. The above chain of equalities indicates that the projection of $\Omega_- f$ onto $R(\Omega_+)$ has norm identical with $\| \Omega_- f \|$. Hence $\Omega_- f \in R(\Omega_+)$ and it follows that $R(\Omega_-) \subseteq R(\Omega_+)$.

To prove (ii) notice that $S^* = \Omega_-^* \Omega_+$ and carry through the same argument as led to (i), with plus and minus interchanged.

The operator S is unitary if and only if $SS^* = S^*S = I$, that is if and only if both S and S^* are isometric. Consequently (iii) follows immediately from (i) and (ii). ∎

A natural consequence of this theorem is

Definition 6.2

The wave operators $\Omega_{\pm}$ are said to be **asymptotically complete** if $R(\Omega_+) = R(\Omega_-)$.

Bearing in mind (6.16) and the argument leading to it we see that asymptotic completeness of the wave operators holds whenever "incoming" (i.e. free as $t \to -\infty$) evolving states correspond to "outgoing" (i.e. free as $t \to +\infty$) evolving states, generated in each case by a suitable free state. Asymptotic completeness is also equivalent to the unitarity of the scattering operator.

6.4 Existence of wave operators

We have spoken a great deal about wave operators but, so far, have not given them a precise definition. We shall now do that and give conditions which ensure their existence.

In Chapter 4 we saw that, given a Hilbert space H and a linear operator $T : H \supseteq D(T) \to H$, the following decomposition of H was available:

$$H = (H_p(T)) \oplus H_c(T) = H_p(T) \oplus H_{ac}(T) \oplus H_{sc}(T)$$

where the various subspaces are introduced in (4.51) and (4.64). We have already remarked that the scattering processes with which we shall be concerned take place in $H_c(T)$, the subspace of continuity of the operator T. Indeed, for most of the problems with which we shall be concerned they take place in $H_{ac}(T)$, a particularly pleasant component of $H_c(T)$ introduced in (4.63). With these remarks in mind the following definition seems quite natural.

Definition 6.3

Let A and A_0 be linear, self adjoint operators on a Hilbert space H. The **wave operators** $\Omega_{\pm} \equiv \Omega_{\pm}(A, A_0)$ associated with the pair $\{A, A_0\}$ are defined by

$$\Omega_{\pm} \equiv \Omega_{\pm}(A, A_0) = s-\lim_{t \to \pm\infty} V^*(t)U(t)P_{ac}(A_0) \tag{6.17}$$

provided these limits exist. Here

$$V^*(t) := \exp(itA), \quad U(t) := \exp(-itA_0)$$

and $P_{ac}(A_0) : H \to H_{ac}(A_0)$ is a projection. We would emphasise that we take the strong limit in (6.17) as introduced in Definition 2.65.

The existence of the wave operators $\Omega_\pm$ can be settled by the following results used either separately or together.

Theorem 6.4

Let D_0 be a subset of $H_{ac}(A_0)$ such that the set D of all finite linear combinations of elements of D_0 is dense in $H_{ac}(A_0)$. If $s-\lim_{t\to\pm\infty} V^*(t)U(t)f$ exists for each $f \in D_0$ then $\Omega_\pm$ exist.

Proof: The statement of the Theorem implies that $\Omega_\pm f$ exists for any $f \in D$. It remains to show that $\Omega_\pm g$ exists for any $g \in H_{ac}(A_0)$. To this end let $\epsilon > 0$ and choose $f \in D$ such that $\| g - f \| < \epsilon/3$. If now we set $W(t) = V^*(t)U(t)$ it will be sufficient to show that $\{W(t)f\}$ is strongly Cauchy as $t \to \pm\infty$ for any $g \in H_{ac}(A_0)$. The existence of $\Omega_\pm g$ will then follow from the completeness of H. Consequently for $s, t > 0$ we have

$$\begin{aligned} \| W(t)g - W(s)g \| &= \| W(t)f - W(s)f - W(t)f + W(s)f + W(t)g - W(s)g \| \\ &\leq \| W(t)f - W(s)f \| + 2 \| g - f \| \end{aligned} \tag{6.18}$$

since $\| V^*(t) \| = \| U(t) \| = I$ for all t

Since $\Omega_\pm f$ exists there must exist a $\tau > 0$ such that

$$\| W(t)f - W(s)f \| < \epsilon/3 \quad \text{whenever } s, t > \tau. \tag{6.19}$$

Combining (6.18) and (6.19) we see that $\{W(t)g\}$ is strongly Cauchy in H for any $g \in H_{ac}(A_0)$ as $t, s \to \pm\infty$. Hence $\Omega_\pm g$ exists as required. ∎

Theorem 6.5

Let $\{W(t) : t \geq a\}$ be a family of linear operators on a Hilbert space H. Let $f \in D(W(t))$ for all t and suppose that $\{W(H)f\}$ is strongly continuously differentiable and that

$$\int_a^\infty \| \frac{d}{d\tau}(W(\tau)f) \| \, d\tau < \infty$$

Then $\{W(t)f\}$ is strongly convergent as $t \to +\infty$.

Proof:

$$\| W(t)f - W(s)f \| = \| \int_s^t \frac{d}{d\tau}(W(\tau)f) d\tau \| \leq \int_s^t \| \frac{d}{d\tau}(W(t)f) \| \, d\tau.$$

the right hand side converges to zero as $s, t \to +\infty$. Hence $\{W(t)f\}$ is strongly convergent. ■

6.5 Selected properties of wave operators

We first introduce the following notion

Definition 6.6

A partial isometry T is an operator which maps a subspace $M_1 \subset H$ isometrically onto another subspace $M_2 \subset H$ and which is zero on $M_1^{\perp}$. The subspace M_1 is the **initial set** of T and M_2 is the range of T.

With the notation of the previous section we now establish a number of results.

Theorem 6.7

(a) The wave operators $\Omega_{\pm}$ are partial isometries on H with initial set $H_{ac}(A_0)$. In particular

$$\Omega_{\pm}^{*}\Omega_{\pm} = P_{ac}(A_0), \quad \Omega_{\pm}P_{ac}(A_0) = \Omega_{\pm}, \quad P_{ac}(A_0)\Omega_{\pm}^{*} = \Omega_{\pm}^{*}$$

(b) Define $F_{\pm} := \Omega_{\pm}\Omega_{\pm}^{*}$. The operator $F_{\pm}$ is a projection onto $R(\Omega_{\pm})$. In particular $F_{\pm}\Omega$

$$F_{\pm}\Omega_{\pm} = \Omega_{\pm}, \quad \Omega_{\pm}^{*}F_{\pm} = \Omega_{\pm}^{*}$$

(c) $\Omega_{\pm}^{*} = s-\lim_{t\to\pm\infty} U^{*}(t)V(t)F_{\pm}$.

Proof: Using the definition of $\Omega_{\pm}$ given in (6.17) and arguing in a similar manner to that which leads to (6.10) we readily establish that $\Omega_{\pm}$ are partial isometries with initial set $H_{ac}(A_0)$. This is emphasised by the observation that if $f, g \in H$ then

$$\begin{aligned}(\Omega_{\pm}^{*}\Omega_{\pm}f, g) &= (\Omega_{\pm}f, \Omega_{\pm}g) = (\Omega_{\pm}P_{ac}(A_0)f, \Omega_{\pm}P_{ac}(A_0)g)\\ &= (P_{ac}(A_0)f, P_{ac}(A_0)g) = (P_{ac}(A_0)f, g).\end{aligned}$$

Hence $\Omega_{\pm}^{*}\Omega_{\pm} = P_{ac}(A_0)$, that is $\Omega_{\pm}^{*}\Omega_{\pm}$ is the identity operator on $H_{ac}(A_0)$. Clearly $\Omega_{\pm} = \Omega_{\pm}P_{ac}(A_0)$ and the final result in (a) follows by taking adjoints.

(b) $F_{\pm}$ is the projection operator introduced in Section 6.3 and the results established there carry over to this case.

(c) Let $f \in H$ then

$$\| U^*(t)V(t)\Omega_\pm f - P_{ac}(A_0)f \| = \| \Omega_\pm f - V^*(t)U(t)P_{ac}(A_0)f \|$$

the right-hand side converges to zero as $t \to \pm\infty$ by definition of $\Omega_\pm$. Consequently we obtain

$$s-\lim_{t\to\pm\infty} U^*(t)V(t)\Omega_\pm = P_{ac}(A_0).$$

Multiply this by $\Omega_\pm^*$ on the right and use (a) to obtain

$$s-\lim_{t\to\pm\infty} U^*(t)V(t)F_\pm = P_{ac}(A_0)\Omega_\pm^* = \Omega_\pm. \qquad \blacksquare$$

Theorem 6.8

Let H be a Hilbert space and $A : H \supseteq D(A) \to H$ a linear, self-adjoint operator. Then

$$(A-\mu I)^{-1} = \begin{cases} i\int_0^\infty e^{i\mu s}e^{-isA}ds, & \operatorname{Im}\mu > 0 \\ -i\int_{-\infty}^0 e^{i\mu s}e^{-isA}ds, & \operatorname{Im}\mu < 0 \end{cases}$$

Proof: Let $\operatorname{Im}\mu > 0$, the proof follows in an analogous manner for $\operatorname{Im}\mu < 0$. Since $\exp(i\mu s)\exp(-isA)$ is strongly continuous in s and

$$\| \exp(i\mu s)\exp(-isA)f \| = \exp(-s\operatorname{Im}\mu) \| f \|$$

then the function $s \to \exp(i\mu s)\exp(-isA)f$ is strongly Riemann integrable on $[0,\infty)$ for each $f \in H$. Consequently $i\displaystyle\int_0^\infty e^{i\mu s}e^{-isA}f ds$ defines an element of H.

Let $f, g \in H$ and use the Spectral Theorem and its associated functional calculus to obtain

$$\begin{aligned}
(g,(A-\mu I)^{-1}f) &= \int_{\mathbf{R}} \frac{1}{\lambda-\mu} d(g,E(\lambda)f) \\
&= \int_{\mathbf{R}} \{i\int_0^\infty e^{i(\mu-\lambda)s}ds\} d(g,E(\lambda)f) \\
&= i\int_0^\infty e^{i\mu s}ds \int_{\mathbf{R}} e^{-i\lambda s}d(g,E(\lambda)f) = i\int_0^\infty e^{i\mu s}(g,e^{-isA}f)ds \\
&= (g, i\int_0^\infty e^{i\mu s}e^{isA}f)
\end{aligned}$$

and the required result follows. ∎

Theorem 6.9

(i) $e^{itA}\Omega_{\pm} = \Omega_{\pm}e^{itA_0}$ for all $t \in \mathbf{R}$ (6.20)

(ii) $(A - \mu I)^{-1}\Omega_{\pm} = \Omega_{\pm}(A_0 - \mu I)^{-1}$ for all $\mu \in \rho(A) \cap \rho(A_0)$ (6.21)

(iii) $E(\lambda, A)\Omega_{\pm} = \Omega_{\pm}E(\lambda, A_0)$ for all $\lambda \in \mathbf{R}$ (6.22)

where $E(\lambda, A)$ and $E(\lambda, A_0)$ are the spectral families of A and A_0 respectively.

Remark: (i) to (iii) are called **intertwining** properties of $\Omega_{\pm}$.

(iv) $F_{\pm}H \subseteq H_{ac}(A)$ (6.23)

Proof: (i) With $V(t) := \exp(-itA)$ and $U(t) := \exp(-itA_0)$ we have

$$\begin{aligned} V^*(t)\Omega_{\pm} &= s-\lim_{\tau\to\pm\infty} V^*(t)V^*(\tau)U(\tau)P_{ac}(A_0) \\ &= s-\lim_{\sigma\to\pm\infty} V^*(\sigma)U(\sigma)U^*(t)P_{ac}(A_0), \quad \sigma := t+\tau. \end{aligned}$$

Direct computation shows that $U^*(t)P_{ac}(A_0) = P_{ac}(A_0)U^*(t)$ and therefore we obtain $V^*(t)\Omega_{\pm} = \Omega_{\pm}U^*(t)$.

(ii) Suppose $\operatorname{Im}\mu > 0$ and use Theorem 6.8 to obtain

$$(A - \mu I)^{-1}\Omega_{\pm} = i\int_0^{\infty} e^{i\mu s}e^{-isA}\Omega_{\pm}ds = i\int_0^{\infty} e^{i\mu s}\Omega_{\pm}e^{-isA_0}ds = \Omega_{\pm}(A_0 - \mu I)^{-1}.$$

Similarly we obtain the result when $\operatorname{Im}\mu < 0$.

(iii) Introduce the notations

$$\begin{aligned} &E(a,b) := E(a,A) - E(b,A), \quad E_0(a,b) := E(a,A_0) - E(b,A_0) \\ &R(t \pm i\epsilon) = [A - (t \pm i\epsilon)I]^{-1}, \quad R_0(t \pm i\epsilon) = [A_0 - (t \pm i\epsilon)]^{-1}. \end{aligned}$$

Stone's Formula, Theorem 4.68, yields, using (ii)

$$\begin{aligned} 2\pi iE(a,b)\Omega_{\pm} &= s-\lim_{\delta\downarrow 0} s-\lim_{\epsilon\downarrow 0}\int_{a+\delta}^{b+\delta}\{R(t+i\epsilon\delta) - R(t-i\epsilon)\}dt\Omega_{\pm} \\ &= s-\lim_{\delta\downarrow 0} s-\lim_{\epsilon\downarrow 0}\int_{a+\delta}^{b+\delta}\Omega_{\pm}\{R_0(t+i\epsilon) - R_0(t-i\epsilon)\}dt \\ &= 2\pi i\Omega_{\pm}E_0(a,b). \end{aligned}$$

Hence

$$E(b,A)\Omega_{\pm} = s-\lim_{a\to-\infty} E(a,b)\Omega_{\pm} = s-\lim_{a\to-\infty} \Omega_{\pm}E_0(a,b) = \Omega_{\pm}E(b,A_0).$$

Remark: It follows from the functional calculus and (iii) that for any bounded, continuous $\phi : \mathbf{R} \to \mathbf{C}$ we have

$$\phi(A)\Omega_{\pm} = \Omega_{\pm}\phi(A_0). \tag{6.24}$$

Further, by taking adjoints $\Omega_{\pm}^{*}\phi(A) = \phi(A_0)\Omega_{\pm}^{*}$.

The adjoints of the results (i) to (iii) follow accordingly.

(iv) Let $f \in F_{+}H$ then $f \in R(\Omega_{+})$ because F_{+} is a projection of H onto $R(\Omega)$ (Theorem 6.7). Consequently, there exists an element $g \in H$ such that $f = \Omega_{+}g$. This implies, due to the unitarity of Ω_{+}, that $g = \Omega_{+}^{*}f \in H_{ac}(A_0)$ (Theorem 6.7(a)). Now recall the notation and arguments leading to (4.57) and obtain, using (6.22)

$$\begin{aligned}\mu_f(I) :=&(f, E(I)f) = (\Omega_{+}g, E(I)\Omega_{+}g) = (g, \Omega_{+}^{*}E(I)\Omega_{+}g) \\ &= (g, E_0(I)\Omega_{+}^{*}\Omega_{+}g) = (g, E_0(I)g) =: \mu_f^0(I).\end{aligned} \tag{6.25}$$

Since $g \in H_{ac}(A_0)$ it follows that $\mu_f^0(I)$ is an absolutely continuous measure and it follows from (6.25) that $\mu_f(I)$ must also be an absolutely continuous measure. Hence $f \in H_{ac}(A)$. A similar result can be obtained using Ω_{-}. ∎

Remark: We notice that in the course of establishing (iv) we have also demonstrated that $\Omega_{\pm}$ map $H_{ac}(A_0)$ into a subspace of $H_{ac}(A)$.

The next result gives an indication of the action of the wave operators on their domains.

Theorem 6.10

If $f \in D(A_0)$ then $\Omega_{\pm}f \in D(A)$ and $A\Omega_{\pm}f = \Omega_{\pm}A_0f$.

Proof:

$$\begin{aligned}\| A\Omega_{\pm}f \|^2 &= \int_{\mathbf{R}} \lambda^2 d \| E(\lambda, A)\Omega_{\pm}f \|^2 = \int_{\mathbf{R}} \lambda^2 d \| \Omega_{\pm}E(\lambda, A_0)f \|^2 \\ &\le \int_{\mathbf{R}} \lambda^2 d \| E(\lambda, A_0)f \|^2 = \| A_0 f \|^2 < \infty.\end{aligned}$$

Hence $\Omega_{\pm}f \in D(A)$.

Furthermore,

$$\| \{\frac{i}{t}(e^{-itA_0} - 1) - A_0\}f \|^2 = \int_{\mathbf{R}} |\frac{i}{t}(e^{-i\lambda t} - 1) - \lambda|^2 d \| E(\lambda, A_0)f \|^2 . \qquad (6.26)$$

Now

$$|\frac{i}{t}(e^{-i\lambda t} - 1) - \lambda| = |\int_0^\lambda (e^{-i\mu t} - 1)d\mu| \leq 2\lambda$$

which indicates that the integrand is uniformly majorised with respect to t by an integrable function. Also the integrand tends to zero as $t \to 0$. Consequently by the Lebesgue Dominated Convergence Theorem we obtain

$$s-\lim_{t\to 0} \{\frac{i}{t}(e^{-itA_0} - 1)f\} = A_0 f.$$

Similarly

$$s-\lim_{t\to 0} \{\frac{i}{t}(e^{-itA} - 1)\Omega_\pm f = A\Omega_\pm f.$$

Since by Theorem 6.9

$$\{\frac{i}{t}(e^{-itA} - 1)\Omega_\pm f\} = \Omega_\pm\{\frac{i}{t}(e^{-itA_0} - 1)f\}$$

we have

$$A\Omega_\pm f = \Omega_\pm A_0 f.$$

■

Theorem 6.11

(i) If either Ω_+ or Ω_- exist then $\sigma_{ac}(A) \supseteq \sigma_{ac}(A_0)$.

(ii) If $F_+ = P_{ac}(A)$ or $F_- = P_{ac}(A)$ then $\sigma_{ac}(A) = \sigma_{ac}(A_0)$

Proof: We notice first that F_+ commutes with $V(t) = \exp(-itA)$ since

$$F_+V(t) = \Omega_+\Omega_+^* V(t) = \Omega_+U^*(t)\Omega_+^* = V(t)\Omega_+\Omega_+^* = V(t)F_+$$

where as usual we have written $U(t) = \exp(-itA_0)$. Similarly we see that F_+ commutes with $E(\lambda, A)$ and hence with A. Consequently, since $F_+H \subseteq H_{ac}(A)$ (Theorem 6.9 (iv)) and $A : H \supseteq D)A) \to A$ we have, in particular, for any $g \in D(A)$ that $F_+Ag \in H_{ac}(A)$. The above commutativity property then yields $AF_+g \in H_{ac}(A)$. Hence

$$\sigma(AF_+) = \sigma_{ac}(AF_+) \subseteq \sigma_{ac}(A). \qquad (6.27)$$

The first equality holds because $F_+ Ag = AF_+ g \in H_{ac}(A)$ for all $g \in D(A)$ implies that the only non-empty component of the spectrum of AF_+ is the absolutely continuous part. The inclusion follows because $D(A) \subseteq H$ and thus as a set $\sigma_{ac}(AF_+)$ cannot exceed $\sigma_{ac}(A)$.

Consequently, to establish (i) it will be sufficient to show, recalling Definition 4.46, that

$$\sigma(AF_+) = \sigma_{ac}(A_0) = \sigma(A_0 P_{ac}(A_0)) \tag{6.28}$$

or equivalently that

$$\rho(AF_+) = \rho(A_0 P_{ac}(A_0)). \tag{6.29}$$

The requirement (6.29) is equivalent to establishing that $(AF_+ - \mu I)^{-1} \in B(F_+ H)$ if and only if $(A_0)P_{ac}(A_0) - \mu I)^{-1} \in B(H_{ac}(A_0))$

where AF_+ is regarded as an operator in $F_+ H$ and $A_0 P_{ac}(A_0)$ as an operator in $H_{ac}(A_0)$.

Suppose that $(AF_+ - \mu I)^{-1} \in B(F_+ H)$ then because $F_+ \Omega_+ = \Omega_+$ we have $(AF_+ - \mu I)^{-1}\Omega = (A - \mu I)^{-1}\Omega_+$.

Now let $f \in D(A_0) \cap H_{ac}(A_0) = D(A_0 P_{ac}(A_0)$. Then using Theorem 6.10 and Theorem 6.7 we can write

$$\begin{aligned} \Omega_+^*(AF_+ - \mu I)^{-1}\Omega_+(A_0 P_{ac}(A_0) - \mu I)f &= \Omega_+^*(AF_+ - \mu I)^{-1}(A - \mu I)F_+\Omega_+ f \\ &= \Omega_+^*\Omega_+ f = f. \end{aligned}$$

Hence $(A_0 P_{ac}(A_0) - \mu I)$ is invertible in $H_{ac}(A_0)$ and its inverse is $\Omega_+^*(AF_+ - \mu I)^{-1}$ which is in $B(H_{ac}(A_0))$.

The converse implication follows by interchanging the roles of Ω_+ and Ω_+^*, of A_0 and A and of $P_{ac}(A_0)$ and F_+. Hence (i) is established.

When $F_+ = P_{ac}(A)$ then $F_+ H = P_{ac}(A)H = H_{ac}(A)$ and, recalling Definition 4.46, we obtain $\sigma(AF_+) = \sigma(AP_{ac}(A)) = \sigma_{ac}(A)$.

The other result follows similarly. ∎

6.6 On the completeness of wave operators

With wave operators defined on $H_{ac}(A_0)$ rather than all of H, as implied in Definition 6.2, we can do a little better than deal with asymptotic completeness. This is due to the fact that, as we have seen in Theorem 6.7 and the remark following Theorem 6.2, $\Omega_\pm$ maps $H_{ac}(A_0)$ isometrically into a subspace of $H_{ac}(A)$.

Definition 6.12

The wave operators $\Omega_\pm$ are said to be **complete** if $R(\Omega_\pm) = H_{ac}(A)$.

Equivalently, recalling Theorem 6.11 (ii), the wave operators are complete if $F_+ = F_- = P_{ac}(A)$. The absolutely continuous parts of A and A_0 are then unitarily equivalent.

Theorem 6.13

The wave operators $\Omega_\pm$ are complete if and only if the limits $s-\lim_{t\to\pm\infty} e^{itA_0}e^{-itA}P_{ac}(A)$ exist.

Proof: For convenience set $V(t) = e^{-itA}$ and $U(t) = e^{-itA_0}$. If the wave operators are complete then by Definition 6.12 and Theorem 6.7(c) we obtain

$$s-\lim_{t\to\pm\infty} U^*(t)V(t)P_{ac}(A) = s-\lim_{t\to\pm\infty} U^*(t)V(t)F_\pm = \Omega_\pm^*.$$

Conversely, let $f \in H_{ac}(A)$ and $g = s-\lim_{t\to\pm\infty} U^*(t)V(t)f$. Then

$$\| f - V^*(t)U(t)g \| = \| U^*(t)V(t)f - g \| \to 0 \text{ as } t \to \pm\infty.$$

Hence $f = \Omega_\pm g$ and therefore, since $f \in H_{ac}(ac)$ was arbitrary,

$$R(\Omega_\pm) = H_{ac}(H).$$

■

Finally, in this chapter we would mention that although we have indicated how questions of existence and completeness of wave operators may be answered we have said little about their structure. Abstract results are available which, for example, express the wave operators, either as the solution of a Lippmann-Schwinger type equation or as a spectral integral. We shall discuss the structure of wave operators for some specific problems in later sections.

Chapter 7
Target scattering

7.1 Introduction

In this chapter we begin an investigation into the distortion of an incident field by a bounded obstacle and of the evolution with time of the resulting field. In contrast to potential scattering, in which the problem is defined on the whole space $\mathbf{R}^n$ and only initial conditions have to be catered for, we shall be concerned here with problems defined over a proper subset $\Gamma \subset \mathbf{R}^n$ and their solutions will be required to satisfy boundary conditions in addition to initial conditions. The resulting initial boundary value problem (IBVP) will be analysed using the techniques outlined in the introductory Chapter 1 and in Chapter 3. Specifically, we shall take as a Free Problem an investigation of the acoustic field in the absence of any obstacles and as a Perturbed Problem a similar study when an obstacle is immersed in the so-called **incident field** generated in the Free Problem.

Before attempting to solve these problems we must decide just what sort of solution we are actually looking for; that is, we must introduce a **solution concept** for each problem. This in turn will indicate suitable function spaces in which to analyse the problems. Next we must settle questions of existence and uniqueness of such solutions. Once this stage has been reached we can then try to obtain more detailed knowledge of such solutions, in our case, their asymptotic behaviour.

7.2 Concerning the incident field

Let $\Gamma \subset \mathbf{R}^3$ denote the unbounded region exterior to the bounded scattering obstacle B. The common boundary between Γ and B is denoted by $\partial\Gamma$ and is assumed to be a smooth, bounded surface. We shall always assume that B is stationary.

The **incident acoustic** field, denoted u_i, satisfies the inhomogeneous d'Alembert equation

$$\left(\frac{\partial^2}{\partial t^2} - \Delta\right) u_i(x,t) = f(x,t), \quad (x,t) \in \mathbf{R}^3 \times \mathbf{R} \tag{7.1}$$

where Δ denotes the Laplacian in $\mathbf{R}^3$ and f is a given function which characterises the acoustic source. We would point out that in (7.1) we have, for convenience, normalised the equilibrium density and the sound speed to unity.

The influence of the source function, f, can be demonstrated by considering, for example, the case when the source, located near a point $x_0 \in \Gamma$, emits a single short pulse of duration τ. In this case we can assume that the support of f satisfies

$$\operatorname{supp} f \subset \{(x,t) : t_0 < t < t_0 + \tau \text{ and } | x - x_0 | < \delta_0\} \tag{7.2}$$

where t_0 and δ_0 are constants. The incident field, u_i, generated by this source satisfies (7.1) and is given by the retarded potential

$$u_i(x,t) = \frac{1}{4\pi} \int_I \frac{f(y, t - | x - y |)}{| x - y |} dy, \quad (x,t) \in \mathbf{R}^3 \times \mathbf{R} \tag{7.3}$$

where $I = \{x, y \in \mathbf{R}^3 : | x - y | \leq \delta_0\}$.

The form of the incident field near the scatterer can be simplified by assuming that the source is at a great distance from B; that is, in the **far field** of B. If we assume that the origin of our coordinate system lies inside B and that there is some constant δ such that $B \subset \{x \in \mathbf{R}^3 : | x | < \delta\}$ then the far field assumption can be written

$$| x_0 | >> \delta_0 + \delta. \tag{7.4}$$

If now we write $x_0 = - | x_0 | \theta_0$ where θ_0 is a unit vector then the term $| x - y |$ can be expressed in the form

$$\begin{aligned} | x - y | &= | x_0 + (y - x_0 - x) | \\ &= \{| x_0 |^2 + 2x_0.(y - x_0 - x) + | y - x_0 - x |^2\}^{\frac{1}{2}} \\ &= | x_0 | \{1 - \theta_0.(y - x_0 - x)/ | x_0 | + O(1/ | x_0 |^2)\}, \end{aligned}$$

as $| x_0 | \to \infty$. Substituting this expression into (7.3) we obtain

$$u_i(x,t) = \frac{s(| x_0 | + x.\theta_0 - t)}{| x_0 |} + O(1/ | x_0 |^2), \quad | x_0 | \to \infty \tag{7.5}$$

where

$$s(\eta) = \frac{1}{4\pi} \int_I f(y, \theta_0.(y - x_0) - \eta) dy, \quad \eta \in \mathbf{R} \tag{7.6}$$

and (7.5) holds uniformly for $|x| \leq \delta$ and $t \in \mathbf{R}$. The quantity $s(\eta)$ is called the **signal waveform (profile)**. We notice that if the error term in (7.5) is neglected then the incident wave is a plane wave. In the following we shall be almost entirely concerned with incident waves which are plane waves.

7.3 A Typical Target Scattering Problem

When a **rigid** obstacle B is immersed in an incident field u_i the standard theory of linear acoustics shows that the resulting **total acoustic field**, u, satisfies

$$\left(\frac{\partial^2}{\partial t^2} - \Delta\right) u(x,t) = 0, \quad (x,t) \in \Gamma \times \mathbf{R} \tag{7.7}$$

$$\frac{\partial u}{\partial n}(x,t) = 0, \quad (x,t) \in \partial\Gamma \times \mathbf{R} \tag{7.8}$$

$$u(x,0) = u_0(x),\ u_t(x,0) = u_1(x), \quad x \in \Gamma \tag{7.9}$$

where $\partial/\partial n$ denotes the derivative in the direction of the unit vector normal to $\partial\Gamma$ drawn from B into Γ. Since Γ is an unbounded region the solution of the IBVP (7.7) to (7.9) will also be required to satisfy certain growth conditions called **radiation conditions**, as $|x| \to \infty$. To give an indication of the nature of these conditions consider, for simplicity, the case when Γ represents a string on $[0,\infty)$; the scattering target in this case being the negative real axis. It is well known that equations of the type (7.7) have solutions which can be written in the form

$$u(x,t) = f(x-t) + g(x+t) \tag{7.10}$$

where f and g are arbitrary functions characterising a wave of constant profile travelling with unit velocity from left to right and right to left respectively. In the particular case when both waves are assumed to have the same time dependence $\exp(-i\omega t)$ then we would expect to be able to write (7.10) in the form

$$u(x,t) = e^{-i\omega t}v_+(x) + e^{-i\omega t}v_-(x). \tag{7.11}$$

Direct substitution of (7.11) into (7.7) shows that the two quantities $v_\pm$ must then satisfy the one equation

$$\left(\frac{d^2}{dx^2} + \omega^2\right) v_\pm(x) = 0. \tag{7.12}$$

However $v_\pm$ are not necessarily the same since

$$v_+(x) = e^{i\omega x} \quad \text{and} \quad v_-(x) = e^{-i\omega x} \tag{7.13}$$

satisfy (7.12). Combining (7.11) and (7.13) we obtain

$$u(x,t) = \exp(-i\omega[t-x]) + \exp(-i\omega[t+x]). \tag{7.14}$$

Thus, recalling the form (7.10), we see that v_+ characterises a wave moving left to right whilst v_- characterises a wave moving right to left, both having the same time dependency $\exp(-i\omega t)$. Equivalently, we say that v_+ is an **outgoing wave** since it is moving away from the origin whilst v_- is an **incoming wave** since it is moving in towards the origin. This particular feature of wave motion can be neatly encapsulated in terms of the **Sommerfeld radiation conditions.**

Definition 7.1

Solutions $v_\pm$ of the equation

$$(\Delta + \omega^2)v_\pm(x) = 0, \quad x \in \mathbf{R}^n$$

are said to satisfy the **Sommerfeld radiation conditions** if and only if

$$\frac{\partial u}{\partial r} \mp i\omega u = o\left(\frac{1}{r}\right), \quad r = |x|, \tag{7.15)(a}$$

$$u(x) = 0\left(\frac{1}{r^{(n-1)/2}}\right) \quad \text{as } r \to \infty. \tag{7.15)(b}$$

The estimates in (7.15)(a), (7.15)(b) are understood to hold uniformly with respect to the direction $x/|x|$.

The estimate (7.15)(a) taken with the minus (plus) sign is called the **Sommerfeld outgoing (incoming) radiation condition.**

With $v_\pm$ defined as in (7.13) it is clear that v_+ is outgoing and v_- is incoming in the sense of Definition 7.1. A derivation of the radiation conditions can be found in the texts cited in the Commentary.

Returning to the IBVP (7.7) to (7.9), we write the total acoustic field, u, in the form

$$u(x,t) = u_i(x,t) + u_s(x,t) \tag{7.16}$$

where u_s, the so-called **scattered field**, is required to satisfy the IBVP

$$\left(\frac{\partial^2}{\partial t^2} - \Delta\right) u_s(x,t) = 0, \quad (x,t) \in \Gamma \times \mathbf{R} \tag{7.17}$$
$$u_s(x,0) = u_0(x) - u_i(x,0), \quad x \in \Gamma$$
$$\frac{\partial u_s}{\partial t}(x,0) = u_1(x) - \frac{\partial u_i}{\partial t}(x,0), \quad x \in \Gamma$$
$$\frac{\partial u_s}{\partial n}(x,t) = -\frac{\partial u_i}{\partial n}(x,t), \quad (x,t) \in \partial\Gamma \times \mathbf{R}.$$

It should be noted that, like u, the scattered field, u_s, is also required to satisfy a radiation condition. Furthermore, we notice that (7.16) exhibits the distortion of the incident field by the target (scatterer).

The following example suggests a means of demonstrating the interplay between u, u_i and u_s.

Example 7.2

A uniform string occupies the region $[0,\infty)$. On this string a wave, characterised by a function g, travels from infinity towards the origin where it is reflected (scattered) by a boundary exhibiting a Neumann condition. The aims are to determine the total displacement of the string at any point x and time t and to prescribe how the displacement evolves with time.

The total displacement of the string, u, satisfies

$$\left\{\frac{\partial^2}{\partial t^2} - c^2\frac{\partial^2}{\partial x^2}\right\} u(x,t) = 0, \quad (x,t) \in (0,\infty) \times \mathbf{R} \tag{7.18}$$
$$u(x,0) = u_0(x), \quad u_t(x,0) = u_1(x), \quad x \in [0,\infty)$$
$$\frac{\partial u}{\partial x}(0,t) = 0, \quad t \in \mathbf{R}$$

where c denotes the constant wave velocity on the string.

It is well known that (7.18) has a solution which can be written in the form

$$u(x,t) = f(x - ct) + g(x + ct), \quad (x,t) \in [0,\infty) \times \mathbf{R} \tag{7.19}$$

where

$$f(\zeta) = 0, \ \zeta < 0 \text{ and } g(\zeta) = 0, \ \zeta > 0. \tag{7.20}$$

The requirements (7.20) are to ensure that the initial conditions can be accommodated. Furthermore, assuming that g is a known function which represents the incident wave then f is determined from the boundary conditions. In this case the boundary condition for (7.18) and (7.19) combine to require that

$$f'(-\zeta)+g'(\zeta)=0, \quad \zeta=ct. \tag{7.21}$$

Integrating we obtain

$$-f(-\zeta)+g(\zeta)=K=\text{constant}. \tag{7.22}$$

Combining (7.20) and (7.22) we find that $K\equiv 0$ and hence (7.19) assumes the form

$$u(x,t)=g(x+ct)+g(-x-ct), \quad (x,t)\in[0,\infty)\times\mathbf{R}. \tag{7.23}$$

The incident wave $g(x+ct)$ is thus seen to be reflected back without any change of form and without any change in sign. The overall effect can be demonstrated graphically in the following manner. Introduce a "virtual" or "image" region $(-\infty,0]$ to complement the "physical" region $[0,\infty)$. On the physical region the incident wave, $g(x+ct)$, travels from $(+\infty)$ towards the origin whilst on the virtual region a mirror image of the incident wave, as dictated by (7.22) and (7.23) namely, $g(-x-ct)$, travels from $(-\infty)$ towards the origin. Both waves are assumed to pass through the boundary at $x=0$, that is the boundary is assumed to be totally transparent as far as these waves are concerned since the boundary conditions have been accommodated in arriving at (7.23). It is clear that for $|t|$ sufficiently large the waves in the physical and virtual regions are well separated whilst for other values of t the two waves will overlap and produce a combined waveform. To fix ideas consider the incident wave to be the triangular or ramp waveform defined by

$$g(\zeta)=\begin{cases}0, & 0\le\zeta\le ct_0\\ \dfrac{\zeta}{c(t_1-t_0)}-\dfrac{t_0}{t_1-t_0}, & ct_0\le\zeta\le ct_1\,.\\ 0, & \zeta\ge ct_1\end{cases} \tag{7.24}$$

If we sketch the progress of this wave and its image in the virtual region then we find that there are four time intervals of particular interest.

(i) $t\le 0$.

Here the waves are well separated and are moving towards each other.

(ii) $t_0\le t\le \frac{1}{2}(t_0+t_1)$.

(iii) $\frac{1}{2}(t_0 + t_1) \leq t \leq t_1$.

In these two regions the waves overlap as they pass through each other. The combined displacement waveform which they produce is complicated.

(iv) $t \geq t_1$.

The waves have passed through each other, are well separated and moving away from each other.

From the resulting diagram we are able to see that in the physical region the total displacement consists of only the incident wave in the interval (i), it consists of a complicated combination of incident and reflected (scattered) waves in the intervals (ii) and (iii) whilst in the interval (iv) it consists only of the reflected wave.

Thus, the diagrams in the intervals (i) and (iv) illustrate the asymptotic behaviour of the total displacement field as $|t| \to \infty$ whilst the diagrams in the intervals (ii) and (iii) give indications of the total field near the scatterer. In developing a scattering theory we try to provide a framework which will cater for both these aspects.

Promising as the approach adopted in Example 7.2 may appear to be it does not lend itself to generalisation. Indeed, the analytical and graphical procedures suggested in Example 7.2 can quickly become intractable. An alternative approach is suggested by the following treatment of the vibrating string problem on a **finite** interval.

Example 7.3

A uniform string occupies the region $(0, \pi) \subset \mathbf{R}$ and its displacement, u, is defined by the IBVP

$$\left\{\frac{\partial^2}{\partial t^2} - \frac{\partial^2}{\partial x^2}\right\} u(x,t) = 0, \quad (x,t) \in (0,\pi) \times \mathbf{R} \tag{7.25}$$

$$u(x,0) = u_0(x) = 1, \quad u_t(x,0) = u_1(x) = 0, \quad x \in (0,\pi) \tag{7.26}$$

$$u(0,t) = 0 = u(\pi,t). \tag{7.27}$$

If this problem is analysed in the Hilbert space $H := L_2(0,\pi)$, then it can be given the operator realisation

$$\left(\frac{\partial^2}{\partial t^2} + A\right) u(x,t) = 0, \quad (x,t) \in (0,\pi) \times \mathbf{R} \tag{7.28}$$

$$u(x,0) = u_0(x) = 1, \quad u_t(x,0) = u_1(x) = 0, x \in (0,\pi) \tag{7.29}$$

where the operator $A : H \supseteq D(A) \to H$ is defined by

$$Au = -u_{xx}, \quad u \in D(A) \tag{7.30}$$

$$D(A) = \{u \in H : u_{xx} \in H,\ u(0,t) = u(\pi,t) = 0\}. \tag{7.31}$$

Now regard $u \in C^2(\mathbf{R}, H)$. The IBVP can then be replaced by the IVP

$$\left(\frac{d^2}{dt^2} + A\right) u(t) = 0 \tag{7.32}$$

$$u(0) = u_0 = 1, \quad u_t(0) = u_1 = 0. \tag{7.33}$$

The problem (7.32), (7.33) has a solution which can be written in the form

$$u(t) = (\cos\ tA^{\frac{1}{2}})u_0 + A^{-\frac{1}{2}}(\sin\ tA^{\frac{1}{2}})u_1 = (\cos tA^{\frac{1}{2}})u_0. \qquad (7.34).$$

To be able to interpret this solution form we first recall the well known result, which it is an easy matter to verify, that the operator A has a complete set of eigenfunctions

$$u_n(x) = \left(\frac{2}{\pi}\right)^{\frac{1}{2}} \sin nx \tag{7.35}$$

with associated eigenvalues

$$\lambda_n = n^2, \quad n \in \mathbf{N}. \tag{7.36}$$

Consequently, for any $u \in D(A)$ and continuous $\phi : \mathbf{R} \to \mathbf{R}$ we have

$$u(x) = \sum_{n=1}^{\infty}(u, u_n)u_n(x), \quad \phi(A)u(x) = \sum_{n=1}^{\infty}\phi(\lambda_n)(u, u_n)u_n(x). \tag{7.37}$$

Therefore, since

$$(u_0, u_n) = \left(\frac{2}{\pi}\right)^{\frac{1}{2}} \int_0^{\pi} \sin nx dx = \begin{cases} \left(\frac{2}{\pi}\right)^{\frac{1}{2}} \cdot \frac{2}{n}, & n \text{ odd} \\ 0, & n \text{ even} \end{cases} \tag{7.38}$$

the spectral decomposition (7.37) indicates that the solution (7.34) can be written

$$u(x,t) = \sum_{n=1}^{\infty}(\cos\ t\lambda_n^{\frac{1}{2}})(u_0, u_n)u_n(x)$$

$$= \frac{4}{\pi}\sum_{n=1}^{\infty}\left(\frac{1}{2n-1}\right)\cos([2n-1]t)\sin([2n-1]x)$$

$$= \frac{2}{\pi}\sum_{n-1}^{\infty}\left(\frac{1}{2n-1}\right)\{\sin([2n-1](x+t))+\sin([2n-1](x-t))\}$$

$$= \frac{1}{2}\{u_0(x+t)+u_0(x-t)\}. \tag{7.39}$$

To obtain (7.39) we simply use the eigenexpansion (7.37). The result (7.39) provides the required interpretation of the solution form (7.34) as it is an expression involving known functions. Furthermore, we notice that (7.39) is precisely the result that would be obtained for this particular problem from the d'Alembert solution (1.5) provided u_0 is an odd function in order to ensure that (7.27) holds.

To summarise the approach adopted in Example 7.3, we first obtained an "$A^{\frac{1}{2}}$-type" solution and then relied on the availability of a Spectral Theorem, typified by the spectral decompositions (7.37), to obtain eigenexpansions which were used to yield an interpretation of the $A^{\frac{1}{2}}$-type solution in terms of given data. The initial conditions (7.26) were displayed explicitly whilst the imposed boundary conditions were absorbed into the definition of the operator A. We would point out however that for more general problems, particularly those involving an unbounded region of interest, the Fourier series appearing in the above example will often have to be replaced by Fourier transforms. This has to be done in order to accommodate the fact that certain exterior problems have no associated eigenfunctions but rather only generalised eigenfunctions.

In principle this approach can be applied to other similar problems in which there appears an operator A_1, say, involving different associated boundary conditions to those for A and which has associated initial conditions v_0, say, rather than u_0.

However, we would emphasise that the great value of the $A^{\frac{1}{2}}$-type solution approach lies in the fact that it provides a mechanism for us to relate u_0 and v_0 in such a way as will ensure that the solutions of the A-problem and the A_1-problem are asymptotically equal as $|t| \to \infty$. This is done in the manner outlined in Chapters 1 and 3. Specifically given v_0, we use associated wave operators to determine those u_0 which will ensure that the A-problem and A_1-problem are asymptotically equal as $|t| \to \infty$. We shall see that the wave operators can be obtained in terms of the eigenfunctions of A and A_1.

7.4 Solution concepts

To provide some motivation for these various notions consider the following differential equation

$$-\frac{du}{dx}(x) = f(x), \quad x \in \Omega := (a,b) \subset \mathbf{R}, \tag{7.40}$$

together with the results outlined in §2.3.

Definition 7.4

If (i) $f \in C(\Omega)$, (ii) $u \in C^1(\Omega)$, (iii) u solves (7.40) then u is said to be a **classical solution** of (7.40).

For convenience at this stage we recall

Definition 7.5

If (i) f is defined on $\mathbf{R}^n$, (ii) $\int_\Omega | f(x) | dx$ exists for every **bounded** $\Omega \subset \mathbf{R}$ then f is said to be **locally integrable** in $\mathbf{R}^n$ and we write $f \in L_{loc}(\mathbf{R}^n)$.

For any classical solution u of (7.40) we always have

$$-\int_\Omega u(x)\frac{d\phi}{dx}(x)dx = \int_\Omega f(x)\phi(x)dx, \quad \text{for all } \phi \in C_0^\infty(\Omega). \tag{7.41}$$

This follows from (7.40) by integration by parts, the integrated terms vanishing since $\phi \in C_0^\infty(\Omega)$. We notice that (7.41) is also meaningful if f and u are only locally integrable, that is, not necessarily continuous, on Ω. This leads naturally to two further solution concepts.

Definition 7.6

If $f \in L_{loc}(\mathbf{R})$ then $u \in L_{loc}(\mathbf{R})$ is a **weak solution** of (7.40) if and only if (7.41) holds for all $\phi \in C_0^\infty(\Omega)$. If u is a weak solution of (7.40) then we say that (7.40) holds in the **weak sense.**

Definition 7.7

If f is a **distribution** which is not necessarily regular, that is not necessarily generated by a locally integrable function, then a distribution u is a solution of (7.40) if any only if

$$-\left(u, \frac{d\phi}{dx}\right) = (f, \phi) \quad \text{for all } \phi \in C_0^\infty(\Omega) \tag{7.42}$$

where (.,.) denotes the "distributional product" introduced in §2.3. We notice that the left hand side of (7.42) defines the distributional derivative $u'(x) = du/dx$. Furthermore, if f is a distribution generated by a locally integrable function and if we seek solutions, u, which are functions then the concept of a **distributional solution** offered by (7.42) reduces to the notion of a weak solution implied by (7.41).

These concepts generalise to $\mathbf{R}^n$. Introduce the linear, partial differential expression of order p defined by

$$L := \sum_{|k|\leq p} a_k(x)D^k \tag{7.43}$$

where

$$\begin{aligned}
&k = (k_1, ..., k_n), \quad |k| = \sum_{m-1}^{n} k_m, \quad k_m \geq 0, \quad m = 1, 2, ...n, \\
&x = (x_1, ..., x_m) \\
&D^k = D_1^{k_1} D_2^{k_2} ... D_n^{k_n}, \quad D_m = \partial/\partial x_m \\
&a_k \in C^\infty(\mathbf{R}^n).
\end{aligned}$$

We have seen that the distribution Lu always exists and is defined by

$$(Lu, \phi) = (u, L^*\phi) \quad \text{for all } \phi \in C_0^\infty(\Omega) \tag{7.44}$$

where here L^* denotes the **formal adjoint** of the differential expression L and is defined by

$$L^*\phi := \sum_{|k|\leq p} (-1)^{|k|} D^k(a_k\phi). \tag{7.45}$$

Corresponding to Definitions 7.6 and 7.7 we now have

Definition 7.8

(i) Let f be a distribution and

$$Lu(x) = f(x), \quad x \in \Omega \subset \mathbf{R}^n. \tag{7.46}$$

A distribution u is a **distributional solution** of (7.46) on Ω if

$$(u, L^*\phi) = (f, \phi), \quad \text{for all } \phi \in C_0^\infty(\Omega). \tag{7.47}$$

(ii) If f is locally integrable then a locally integrable function u which satisfies (7.47) is a **weak solution** of (7.46).

(iii) If $f \in C(\Omega)$ and $u \in C^p(\Omega)$ satisfies (7.46) at all points $x \in \Omega \subset \mathbf{R}^n$, then u is **a classical solution** of (7.46).

Another solution concept, and one which will be of particular interest to us when developing scattering theories, is one which is directly associated with the energy in the wave field. By analogy with the energy of a string performing small transverse vibrations we introduce the following:

Definition 7.9

The wave energy, $E(u, \Omega, t)$, in a domain $\Omega \subset \mathbf{R}^n$ at time t of a solution $u(x,t)$ of an IBVP associated with the wave equation in $\mathbf{R}^n \times \mathbf{R}$ is defined by

$$E(u, \Omega, t) = \int_{\Omega} | u_t(x,t) |^2 + \sum_{m=1}^{n} | D_m u(x,t) |^2 \} dx \tag{7.48}$$

where $x = (x_1, ..., x_n)$ and $dx = dx_1 dx_2 ... dx_n$ denotes the volume element in $\mathbf{R}^n$.

Comparing (7.48) with the corresponding expression for a vibrating string we call the first and second terms on the right-hand side the kinetic and potential energy terms respectively.

We shall be interested in solutions of IBVP for which energy is conserved in the sense that

$$E(u, \Omega, t) = E(u, \Omega, 0) = \text{constant}, \quad t \in \mathbf{R}. \tag{7.49}$$

The constant in (7.49) could be either finite or infinite. We shall only be interested in solutions for which this constant is finite, that is, in **solutions with finite energy**.

7.5 Concerning existence and uniqueness of solutions

We have seen that an IBVP for the wave equation can be replaced by the following equivalent IVP for an ordinary differential equation: Determine an element $u \in C^2(\mathbf{R}, D(A)) \cap C(\mathbf{R}, H)$ satisfying

$$u'' + Au = 0, \quad u(0) = u_0, \quad u_t(0) = u_1 \qquad \mathbf{P1}$$

where $A : H \supseteq D(A) \to H$ is defined by

$$Au = -\Delta u \quad \text{for all } u \in D(A)$$

$$D(A) := \{u \in H : \Delta u \in H \text{ and } u \in (bc)\}$$

and $u \in (bc)$ denotes that the solution u is required to satisfy certain conditions imposed on the boundary of the region of interest.

Assume, for the moment, that **P1** is well posed. This, as we have seen in Chapter 5, then immediately settles questions of existence and uniqueness of solution. Therefore, knowing that a solution of **P1** exists we express it in the form

$$u(t) = (\cos\ tA^{\frac{1}{2}})u_0 + A^{-\frac{1}{2}}(sin\ tA^{\frac{1}{2}})u_1 \tag{7.50}$$

which is meaningful provided (i) $u_0, u_1 \in D(A^{\frac{1}{2}})$ and (ii) A is self adjoint, in order that the Spectral Theorem is available for interpreting (7.50).

To settle the well-posedness of **P1** we reduce it to the first order system

$$\psi'(t) - iM\psi(t) = 0, \quad \psi(0) = \psi_0 \qquad \textbf{P2}$$

where

$$\psi(t) = \begin{bmatrix} u \\ u' \end{bmatrix}(t), \quad \psi_0 = \begin{bmatrix} u_0 \\ u_1 \end{bmatrix}, \quad M = i\begin{bmatrix} 0 & -I \\ A & 0 \end{bmatrix} \tag{7.51}$$

and

$$\psi = (\psi_1, \psi_2) \in H \times H, \quad M : H \times H \supseteq D(M) \to H \times H. \tag{7.52}$$

We shall give below a precise definition of $D(M)$ which will be in keeping with our interest in solutions with finite energy.

If (iM) is a self-adjoint operator then by Theorem 5.10 we see that **P2** is well posed and has a unique solution given by

$$\psi(t) = e^{itM}\psi_0. \tag{7.53}$$

The assumed self-adjointness of (iM) means that the Spectral Theorem is available and that we can write

$$\begin{aligned} e^{itM} &= \sum_{n=0}^{\infty} \frac{(itM)^n}{n!} = \left\{ \sum_{n \text{ even}} + \sum_{n \text{ odd}} \right\} \frac{(itM)^n}{n!} \\ &= \{I - \frac{t^2M^2}{2!} + \frac{t^4M^4}{4!} - \ \dots\ \} + i\{tM - \frac{t^3M^3}{3!} + \frac{t^5M^5}{5!} - \ \dots\ \} \end{aligned}$$

Now, using

$$M = i\begin{bmatrix} 0 & -1 \\ A & 0 \end{bmatrix}, \quad M^2 = A\begin{bmatrix} I & 0 \\ 0 & I \end{bmatrix}$$

and recalling the series expansions of $\cos x$ and $\sin x$ we obtain

$$\begin{aligned} e^{itM} &= \{I - \frac{t^2 A}{2!} + \frac{t^4 A^2}{4!} - \dots\}\begin{bmatrix} I & 0 \\ 0 & I \end{bmatrix} + \\ &- A^{-\frac{1}{2}}\{tA^{\frac{1}{2}} - \frac{t^3 A^{\frac{3}{2}}}{3!} + \frac{t^5 A^{\frac{5}{2}}}{5!} - \dots\}\begin{bmatrix} 0 & -I \\ A & 0 \end{bmatrix} \\ &= (\cos\ tA^{\frac{1}{2}})\begin{bmatrix} I & 0 \\ 0 & I \end{bmatrix} - A^{-\frac{1}{2}}(\sin\ tA^{\frac{1}{2}})\begin{bmatrix} 0 & -I \\ A & 0 \end{bmatrix}. \end{aligned} \tag{7.54}$$

Substituting (7.54) into (7.53) we obtain as the first component of $\psi(t)$

$$u(t) = (\cos\ tA^{\frac{1}{2}})u_0 + A^{-\frac{1}{2}}(\sin\ tA^{\frac{1}{2}})u_1 \tag{7.55}$$

and thus we have recovered the solution (7.50). We notice from (7.53) and (7.54) that u, given by (7.55), is well defined provided $u_0 \in D(A)$ and $u_1 \in D(A^{\frac{1}{2}})$. Furthermore, it is well known that if a linear operator $T : H \supseteq D(T) \to H$ is positive, self-adjoint then there exists a unique square root $T^{\frac{1}{2}}$ which is also positive and self-adjoint.

These various results can be summarised in the following.

Theorem 7.10

Let H be a Hilbert space and $A : H \supseteq D(A) \to H$ a positive, self-adjoint, linear operator. If the operator $M : H \times H \supseteq D(M) \to H \times H$ defined by $M = i\begin{bmatrix} 0 & -I \\ A & 0 \end{bmatrix}$ is self-adjoint then (iM) generates a C_0-group $\{U(t),\ t \in \mathbf{R}\}$ defined by

$$U(t) = e^{itM} = (\cos\ tA^{\frac{1}{2}})\begin{bmatrix} I & 0 \\ 0 & I \end{bmatrix} - A^{-\frac{1}{2}}(\sin\ tA^{\frac{1}{2}})\begin{bmatrix} 0 & -1 \\ A & 0 \end{bmatrix}. \tag{7.56}$$

Consequently, the IBVP

$$u'' + Au = 0, \quad u(0) \in D(A), \quad u'(0) = u_1 \in D(A^{\frac{1}{2}})$$

is well posed.

A rigorous proof of this theorem can be found in the references cited in the Commentary.

Thus, when developing a scattering theory for an IBVP associated with **P1** we see that existence and uniqueness results can be established once we show that the operators $A : H \supseteq D(A) \to H$ and $M : H \times H \supseteq D(M) \to H \times H$ are positive, self-adjoint operators on their respective spaces. To illustrate how this can be done we shall confine attention to a free problem, that is one posed in all of $\mathbf{R}^n$. We shall return to the more general problem involving a target in the next Chapter.

The self-adjointness of M can be conveniently established in the **energy** space, $H_E(\mathbf{R}^n)$, defined by

$$H_E(\mathbf{R}^n) := H_D(\mathbf{R}^n) \times L_2(\mathbf{R}^n) \tag{7.57}$$

where $H_D(\mathbf{R}^n)$ is the closure of $C^\infty(\mathbf{R}^n)$ with respect to the norm

$$\| f \| H_D := \{\int_{\mathbf{R}^n} | \nabla f(x) |^2 \, dx\}^{\frac{1}{2}}. \tag{7.58}$$

The norm in H_E is defined by

$$\| f \|_E^2 = \int_{\mathbf{R}^n} \{| \nabla f_1(x) |^2 + | f_2(x) |^2\} dx \tag{7.59}$$

for all $f = (f_1, f_2) \in H_D(\mathbf{R}^n) \times L_2(\mathbf{R}^n) = H_E(\mathbf{R}^n)$. The associated inner product is

$$(f, g)_E = (\nabla f_1, \nabla g_1) + (f_2, g_2), \quad f, g \in H_E \tag{7.60}$$

where $(.,.)$ denotes the usual $L_2(\mathbf{R}^n)$ inner product.

For $M : H_E(\mathbf{R}^n) \supseteq D(M) \to H_E(\mathbf{R}^n)$ defined by $M = i \begin{bmatrix} 0 & -I \\ A & 0 \end{bmatrix}$ we say that $f = (f_1, f_2)$ is an element of $D(M)$ provided $Mf \in H_E(\mathbf{R}^n)$, that is provided

$$\begin{bmatrix} 0 & -I \\ A & 0 \end{bmatrix} \begin{bmatrix} f_1 \\ f_2 \end{bmatrix} = \begin{bmatrix} -f_2 \\ Af_1 \end{bmatrix} \in H_E(\mathbf{R}^n).$$

Consequently we define

$$D(M) = \{f = (f_1, f_2) \in H_E(\mathbf{R}^n) : f_2 \in H_D(\mathbf{R}^n), \; Af_1 \in L_2(\mathbf{R}^n)\}. \tag{7.61}$$

To prove that M is self-adjoint on $H_E(\mathbf{R}^n)$ we see from Theorem 2.98 that it is sufficient to establish that $R(M \pm iI) = H_E(\mathbf{R}^n)$. To this end we first show that it is symmetric on H_E. For any $f = (f_1, f_2) \in H_E(\mathbf{R}^n)$ and any $\phi = (\phi_1, \phi_2) \in C_0^\infty(\mathbf{R}^n) \times C_0^\infty(\mathbf{R}^n)$ we use (7.60) and one integration by parts using Green's Theorem, remembering that all integrated terms vanish by the properties of ϕ, to obtain

$$(Mf, g)_E = \left(i \begin{bmatrix} -f_2 \\ Af_1 \end{bmatrix}, \begin{bmatrix} g_1 \\ g_2 \end{bmatrix} \right)_E = i(-\nabla f_2, \nabla g_1) + i(\nabla f_1, \nabla g_2)$$

$$(f, Mg)_E = \left(\begin{bmatrix} f_1 \\ f_2 \end{bmatrix}, \begin{bmatrix} -g_2 \\ Ag_1 \end{bmatrix} \right)_E = i(\nabla f_1, \nabla g_2) + i(-\nabla f_2, \nabla g_1).$$

Therefore since $C_0^\infty(\mathbf{R}^n)$ is dense in $H_D(\mathbf{R}^n)$ and $L_2(\mathbf{R}^n)$ we conclude that these relations hold on $H_E(\mathbf{R}^n)$. Hence M is symmetric on $H_E(\mathbf{R}^n)$.

Since M is symmetric on $H_E(\mathbf{R}^n)$ Theorem 2.96 (vii) indicates that it is also closable there. Indeed we can actually show that it is a closed operator.

Let $\overline{M}$ denote the closure of M. We shall show that $D(\overline{M}) = D(M)$ and hence that M is closed.

Recall that $D(\overline{M})$ is the closure of $D(M)$ with respect to the graph norm, $\| \cdot \|_M$, which in this case is

$$\| f \|_M^2 = \| f \|_E^2 + \| Mf \|_E^2 \quad \text{for all } f \in D(M). \tag{7.62}$$

Since $Mf = \begin{bmatrix} -f_2 \\ Af_1 \end{bmatrix}$ for $f = (f_1, f_2) \in D(M)$ then (7.62) can be written

$$\| f \|_M^2 = \| f_1 \|_D^2 + \| f_2 \|^2 + \| -if_2 \|_D^2 + \| Af_1 \|^2 . \tag{7.63}$$

Now $f \in D(M) \subset H_E(\mathbf{R}^n)$ obviously implies that $f \in D(\overline{M})$ and thus $D(M) \subseteq D(\overline{M})$. On the other hand the subset $D(\overline{M}) \subset H_E(\mathbf{R}^n)$ consisting of those elements $f = (f_1, f_2) \in H_E(\mathbf{R}^n)$ which have a finite norm $\| f \|_M$ as defined by (7.63) must be those elements for which $f_2 \in H_D(\mathbf{R}^n)$ and $Af_1 \in L_2(\mathbf{R}^n)$. That is, $f \in D(\overline{M})$ implies that $f \in D(M)$ and hence $D(\overline{M}) \subseteq D(M) \subset H_E(\mathbf{R}^n)$. Combining these two inclusions we obtain $D(M) = D(\overline{M})$ and hence M is a closed operator in $H_E(\mathbf{R}^n)$.

Next we show that $R(M \pm iI)$ is closed in $H_E(\mathbf{R}^n)$. To see this first notice that

$$\| (M \pm iI)\phi \|_E^2 = \| M\phi \|_E^2 \mp i(M\phi, \phi)_E \pm i(\phi, M\phi)_E + \| \phi \|_E^2 .$$

Since M is symmetric the second and third terms on the right hand side of this expression cancel and we obtain

$$\| (M \pm iI)\phi \|_E \geq \| \phi \|_E \quad \text{for all } \phi \in D(M). \tag{7.64}$$

From (7.64) and Theorem 2.91(a) we can conclude that $R(M \pm iI)$ are closed if and only if $(M \pm iI)$ are closed operators. Since M is a closed operator on $H_E(\mathbf{R}^n)$ it follows that the operators $(M \pm iI)$ are also closed operators on $H_E(\mathbf{R}^n)$. Hence $R(M \pm iI)$ are closed in $H_E(\mathbf{R}^n)$.

Finally, we shall show that the ranges $R(M \pm iI)$ are dense in $H_E(\mathbf{R}^n)$. To prove this consider the equations

$$(M \pm iI)\phi = h \tag{7.65}$$

where $\phi = (\phi_1, \phi_2)$, $h = (h_1, h_2)$ and $\phi_k, h_k \in C_0^\infty(\mathbf{R}^n)$, $k = 1, 2$.

Writing (7.65) in component form we obtain the equations

$$\mp \phi_1 + \phi_2 = ih_1 \tag{7.66}$$

$$\Delta\phi_1 \mp \phi_2 = ih_2 \tag{7.67}$$

from which it follows that

$$\Delta\phi_1 - \phi_1 = i(h_2 \pm h_1). \tag{7.68}$$

Taking Fourier transforms throughout (7.68) we obtain

$$F((\Delta - I)\phi_1)(\xi) = (- \mid \xi^2 \mid +1)\hat{\phi}_1(\xi) = i(\hat{h}_2(\xi) \pm \hat{h}_1(\xi)).$$

Consequently, solving this equation for $\hat{\phi}_1$ and using the Convolution Theorem for Fourier transforms (recall, $F(\theta_1 * \theta_1)(\xi) = \hat{\theta}_1(\xi)\hat{\theta}_2(\xi))$ we can obtain ϕ_1. We then obtain ϕ_2 from (7.66). Thus we have shown that for any $h_1, h_2 \in C_0^\infty(\mathbf{R}^n)$ we can always find $\phi_1, \phi_2 \in C_0^\infty(\mathbf{R}^n)$ such that (7.65) holds. Consequently we must have

$$C_0^\infty(\mathbf{R}^n) \times C_0^\infty(\mathbf{R}^n) \subset R(M \pm iI).$$

Therefore, if we recall the definition of $H_D(\mathbf{R}^n)$ and also the fact that $C_0^\infty(\mathbf{R}^n)$ is dense in $L_2(\mathbf{R}^n)$ then we can conclude that $R(M \pm iI)$ must be dense in $H_E(\mathbf{R}^n)$.

Since we already know that the $R(M \pm iI)$ are closed in $H_E(\mathbf{R}^n)$ it follows immediately that

$$R(M \pm iI) = H_E(\mathbf{R}^n).$$

Hence M is self-adjoint in $H_E(\mathbf{R}^n)$.

Chapter 8
A scattering theory

8.1 Introduction

Acoustic wave propagation phenomena are known to model many evolutionary processes in the applied sciences. In the previous Chapters we have discussed Initial Boundary Value Problems (IBVP) associated with such phenomena and now have a reasonable confidence that we can settle questions of existence and uniqueness of solution to such problems. However, it must be emphasised that being able to obtain existence and uniqueness results for these problems provide only a first step towards understanding the actual structure of solutions; so far, we have no more than superficial information in this connection.

In this Chapter we introduce a method of spectral and asymptotic analysis, based on expansions with respect to generalised eigenfunctions, which can treat a wide range of physically interesting wave propagation phenomena. To illustrate the method we shall restrict attention to acoustic wave propagation problems and summarise the pioneering works of C.H. Wilcox [85] in this connection. In doing this we shall be mainly concerned with stating the principal results and with indicating the various steps leading up to them. Many of the results are of a technical nature and their proofs, whilst frequently elegant, are often quite lengthy. For this reason proofs will be omitted but reference made to them in the Commentary to this Chapter.

We take as the Free Problem (FP) the investigation of acoustic waves in $\mathbf{R}^3$. Solutions are constructed using the Plancherel theory of Fourier transforms in $L_2(\mathbf{R}^3)$ and the asymptotic behaviour of solutions having finite energy is obtained. For the Perturbed Problem (PP), which arises when an obstacle is introduced into the "incident field" discussed in the FP, similar results can be obtained provided the plane wave kernels used in generating Fourier transforms for the FP are replaced by the distorted plane waves, introduced by Ikebe [29], to generate so-called generalised Fourier transforms. Asymptotic results can be obtained which indicate that the wave profile for the FP and for the PP agree as $t \to \infty$. We shall see that the associated wave operators can then be obtained in terms of the Fourier transforms used in the FP and the generalised Fourier transforms used in the PP.

8.2 A free problem

The generic IBVP with which we shall be concerned has the form

$$\left(\frac{\partial^2}{\partial t^2} - \Delta\right) w(x,t) = 0, \quad (x,t) \in \Omega \times \mathbf{R} \tag{8.1}$$

$$w(x,0) = f(x), \quad w_t(x,0) = g(x), x \in \Omega \tag{8.2}$$

$$w(x,t) \in (bc), \quad (x,t) \in \partial\Omega \times \mathbf{R} \tag{8.3}$$

where Ω is an open region in $\mathbf{R}^3$ with boundary $\partial\Omega$. The quantities f, g represent initial data and (bc) indicates that solutions of (8.1) are required to satisfy certain, as yet, unspecified boundary conditions on $\partial\Omega$.

For the FP we take $\Omega = \mathbf{R}^3$ and introduce the operator

$$A_0 : L_2(\mathbf{R}^3) \to L_2(\mathbf{R}^3) =: H(\mathbf{R}^3) \tag{8.4}$$

$$A_0 u_0 = -\Delta u_0, \quad u \in D(A_0)$$

$$D(A_0) = \{u \in H(\mathbf{R}^3) : \Delta u_0 \in H(\mathbf{R}^3)\}.$$

The first result which can be obtained is

Theorem 8.1

(i) A_0 is a self-adjoint, non-negative operator on $H(\mathbf{R}^3)$.

(ii) A_0 has a unique, non-negative, square root $A_0^{\frac{1}{2}}$ with domain

$$D(A_0^{\frac{1}{2}}) = \{u \in H(\mathbf{R}^3) : D^\alpha u \in H(\mathbf{R}^3), \mid \alpha \mid \leq 1\}$$

where $\alpha = (\alpha_1, ...\alpha_n)$ is a multi-index with α_j non-negative integers, $j = 1, 2, ...n$.

We now interpret (8.1) as

$$\left(\frac{d^2}{dt^2} + A_0\right) u_0(t) = 0, \quad t \in \mathbf{R} \tag{8.5}$$

where the notation emphasises that $u_0(x,t)$, the unknown for (8.1), is now understood to be an $H(\mathbf{R}^3)$ valued function of $t \in \mathbf{R}$. The initial values corresponding to (8.2) are consequently of the form

$$u_0(0) = \phi_0, \quad u_{0t}(0) = \phi_1 \tag{8.6}$$

where it is assumed that $\phi_0, \phi_1 \in H(\mathbf{R}^3)$.

A solution of the Initial Value Problem (IVP) (8.5), (8.6) can be written in the form

$$u_0(t) = (\cos tA_0^{\frac{1}{2}})\phi_0 + A_0^{-\frac{1}{2}}(\sin tA_0^{\frac{1}{2}})\phi_1. \tag{8.7}$$

If, further, ϕ_0, ϕ_1 are real valued functions such that $\phi_0 \in H(\mathbf{R}^3)$ and $\phi_1 \in D(A_0^{\frac{1}{2}})$ and if we define on $H(\mathbf{R}^3)$

$$h_0 := \phi_0 + iA_0^{-\frac{1}{2}}\phi_1 \tag{8.8}$$

then the solution (8.7) can be expressed in the form

$$u_0(t) \equiv u_0(.,t) = Re\{v_0(.,t)\} \tag{8.9}$$

where

$$v_0(t) \equiv v_0(.,t) := \exp(-itA_0^{\frac{1}{2}})h_0 \tag{8.10}$$

is the complex-valued solution in $H(\mathbf{R}^3)$ of (8.5), (8.6).

Since we know, by Theorem 8.1, that A_0 is self-adjoint then the Spectral Theorem is available for interpreting coefficients involving $A_0^{\frac{1}{2}}$. Specifically, if $\{E_0(\lambda)\}$ denotes the spectral family of A_0 then we have the spectral representations

$$A_0 = \int_0^\infty \lambda dE_0(\lambda) \tag{8.11}$$

$$\Phi(A_0) = \int_0^\infty \Phi(\lambda) dE_0(\lambda) \tag{8.12}$$

where Φ is a bounded, Lebesgue measurable function of λ. Consequently, $A_0^{\frac{1}{2}}$ is interpreted by setting $\Phi(\lambda) = \lambda^{\frac{1}{2}}$ in (8.12). The remaining coefficients in (8.7) and (8.8) are interpreted similarly.

However, a practical difficulty centred on (8.11) and (8.12) concerns the actual determination of the spectral family $\{E_0(\lambda)\}$. In the case of the FP the situation can be eased by introducing Fourier transforms in $H(\mathbf{R}^3)$, the Plancherel theory of which indicates that for any $f \in H(\mathbf{R}^3)$ the following limits exist

$$F(f)(p) := \hat{f}(p) = \lim_{R\to\infty} \frac{1}{(2\pi)^{\frac{3}{2}}} \int_{|x|\leq R} e^{-ix.p} f(x)dx \tag{8.13}$$

$$f(x) = F^{-1}(\hat{f})(x) = \lim_{R\to\infty} \frac{1}{(2\pi)^{\frac{3}{2}}} \int_{|p|\leq R} e^{ix.p} \hat{f}(p)dp. \tag{8.14}$$

It can also then be shown that for any bounded Lebesgue measurable function Φ we have

$$\Phi(A_0)f(x) = \lim_{R\to\infty} \frac{1}{(2\pi)^{\frac{3}{2}}} \int_{|p|\leq R} e^{ixp} \Phi(|\, p\,|^2)\hat{f}(p)dp. \tag{8.15}$$

We would emphasise that the limits in (8.13) to (8.15) must be taken in the $H(\mathbf{R}^3)$ sense. Furthermore, the operator $F : H(\mathbf{R}^3) \to H(\mathbf{R}^3)$ is unitary and consequently we shall sometimes write $F^{-1} = F^*$.

We now notice that

$$w_0(x,p) := \frac{1}{(2\pi)^{\frac{3}{2}}} \exp(ix.p), \quad p \in R^3 \tag{8.16}$$

satisfies the Helmholtz equation

$$(\Delta + |\, p\,|^2)w_0(x,p) = 0 \quad \text{for all } x, p \in \mathbf{R}^3. \tag{8.17}$$

Thus w_0 might be thought to be an ordinary eigenfunction of A_0 with associated spectral parameter $|\, p\,|^2$. In fact w_0 is a **generalised** eigenfunction of A_0 since, as direct calculation shows, $w_0 \notin H(\mathbf{R}^3)$. Nevertheless, a generalisation of the spectral decomposition of an operator used in Example 7.3 can be obtained in terms of these generalised eigenfunctions. Specifically, we re-write (8.13) to (8.15) using (8.16) to obtain

$$F(f)(p) := \hat{f}(p) = \lim_{R\to\infty} \int_{|x|\leq R} \overline{w_0(x,p)} f(x)dx \tag{8.18}$$

$$f(x) = F^{-1}(\hat{f})(x) = \lim_{R\to\infty} \int_{|p|\leq R} w_0(x,p)\hat{f}(p)dp \tag{8.19}$$

$$\Phi(A_0)f(x) = \lim_{R\to\infty} \int_{|p|\leq R} w_0(x,p)\Phi(|\, p\,|^2)\hat{f}(p)dp \tag{8.20}$$

where as before the limits have to be taken in an $H(\mathbf{R}^3)$ sense. It is interesting to note that (8.20) can also be written in the form

$$F\Phi(A_0)f(p) = \Phi(|\, p\,|^2)\hat{f}(p). \tag{8.20a}$$

Furthermore, if we now compare (8.20) and (8.12) then we might expect also to obtain information about $\{E_0(\lambda)\}$ the spectral family of A_0. For instance a direct comparison of (8.20) and (8.12) would seem to suggest that, in some sense,

$$dE_0(\mu)f(x) = w_0(x,p)\hat{f}(p)dp. \tag{8.21}$$

To make this more precise we first recall that $E_0(\mu)$ satisfies

$$E_0(\mu) = \begin{cases} 1, & \mu \geq 0 \\ 0, & \mu < 0 \end{cases} \tag{8.22}$$

Consequently, $E_0(\mu)$ has the property of the Heaviside unit function $H(\tau)$. Thus, if in (8.20) we take $\Phi(\lambda) = H(\mu - \lambda)$ then we obtain

$$E_0(\mu)f(x) = \lim_{R\to\infty} \int_{|p|\leq R} w_0(x,p)H(\mu - \mid p\mid^2)\hat{f}(p)dp$$

from which it follows that

$$E_0(\mu)f(x) = \begin{cases} \int_{|p|\leq\sqrt{\mu}} w_0(x,p)\hat{f}(p)dp, & \mu \geq 0 \\ 0, & \mu < 0 \end{cases} \tag{8.23}$$

Differentiating (8.23) we recover (8.21).

These various results enable us to interpret the coefficients appearing in the solution forms (8.7) and (8.10). Specifically the results imply that the **wave function** $v_0(x,t)$ defined by (8.10) has the representation

$$v_0(x,t) = \int_{\mathbf{R}^3} w_0(x,p)\exp(-it\mid p\mid)\hat{h}_0(p)dp. \tag{8.24}$$

Notice that the improper integral in (8.24) must be interpreted in the $H(\mathbf{R}^3)$ limit sense used in results such as (8.20). We shall frequently use the more convenient notation of (8.24) with this understanding.

In (8.24) we notice that

$$w_0(x,p)\exp(-it\mid p\mid) = \frac{1}{(2\pi)^{\frac{3}{2}}}\exp(i(x.p - t\mid p\mid)) \tag{8.25}$$

are solutions of the wave equation (8.1) which represent plane waves propagating in the direction of the vector p. Therefore, the wave function given by (8.24) is a representation of an acoustic wave in terms of the elementary waves (8.25).

Now that we can interpret the terms in the various representations we can summarise the results which can be obtained so far concerning the solutions of the FP in the form

Theorem 8.2

Let $\phi_0 \in H(\mathbf{R}^3)$ and $\phi_1 \in D(A_0^{\frac{1}{2}})$ be real valued functions and define h_0 by (8.8). Then the solution in $H(\mathbf{R}^3)$ of the FP, defined by (8.7), satisfies

$$u_0(x,t) = \mathrm{Re}(v_0(x,t))$$

where $v_0(x,t)$ is the complex valued solution in $H(\mathbf{R}^3)$ of the wave equation (8.1) defined by (8.10).

This Theorem suggests that an asymptotic analysis as $t \to \infty$ of solutions to the FP should be centred on the spectral integral (8.24). To indicate how this can be achieved consider the special case when $\hat{h}_0(p)$ in (8.24) has the properties

$$\begin{aligned} &\hat{h}_0 \in \{f \in C_0^\infty(\mathbf{R}^3) : f(p) \equiv 0, \mid p \mid \leq a, \quad a > 0\} \\ &\operatorname{supp} \hat{h}_0 \subset \{p : 0 < a \leq \mid p \mid \leq b\}. \end{aligned} \tag{8.26}$$

In this case the spectral integral (8.24) converges in $H(\mathbf{R}^3)$ pointwise to v_0 and

$$v_0(x,t) = \frac{1}{(2\pi)^{\frac{3}{2}}} \int_D \exp\{i(x.p - t \mid p \mid)\}\hat{h}_0(p)dp \tag{8.27}$$

where D denotes the interval $a \leq \mid p \mid \leq b$.

To determine the asymptotic properties of $v_0(.,t) \in H(\mathbf{R}^3)$ as $t \to \infty$ introduce polar coordinates in the form

$$p = \rho\omega,\ \rho \geq 0,\ \omega \in S^2,\ dp = \rho^2 d\rho d\omega \tag{8.28}$$

where S^2 denotes the unit sphere in $\mathbf{R}^3$ with centre the origin and $d\omega$ is the element of surface area of S^2. The integral (8.27) now assumes the form

$$v_0(x,t) = \frac{1}{(2\pi)^{\frac{3}{2}}} \int_a^b e^{-it\rho} V_0(x,\rho)\rho^2 d\rho \tag{8.29}$$

where

$$V_0(x,\rho) = \int_{S^2} e^{i\rho x.\omega}\hat{h}_0(\rho\omega)dw. \tag{8.30}$$

The method of stationary phase applied to (8.30) with $x = r\theta$, $r > 0$ and $\theta \in S^2$ implies that if

$$V_0(x,\rho) = \frac{2\pi i}{\rho r}\{-e^{i\rho r}\hat{h}_0(\rho\theta) + e^{-i\rho r}\hat{h}_0(-\rho\theta)\} + q_0(x,\rho) \tag{8.31}$$

then there exists a constant $M_0 = M_0(\hat{h}_0)$ such that $| q_0(x,\rho) | \leq \frac{M_0}{r^2}$, for all $r > 0$, $a \leq \rho \leq b$, $\theta \in S^2$. Substituting (8.31) into (8.29) we obtain

$$v_0(x,t) = \frac{G(r-t,\theta)}{r} + \frac{G'(r+t,\theta)}{r} + q_1(x,t) \tag{8.32}$$

where

$$G(\tau,\theta) = \frac{1}{(2\pi)^{\frac{1}{2}}}\int_a^b e^{i\tau\rho}\hat{h}_0(\rho\theta)(-i\rho)d\rho \tag{8.33}$$

$$G'(\tau,\theta) = \frac{1}{(2\pi)^{\frac{1}{2}}}\int_{-b}^{-a} e^{i\tau\rho}\hat{h}_0(\rho\theta)(-i\rho)d\rho \tag{8.34}$$

and q_1 is a quantity which, on using the above estimate for q_0, satisfies

$| q_1(x,t) | \leq \frac{M_1}{r^2}$ for all $r > 0$, $t \in \mathbf{R}$, $\theta \in S^2$ with $M_1 = M_1(\hat{h}_0) = (2\pi)^{-\frac{3}{2}}(b^3 - a^3)M_0(\hat{h}_0)/3$

The result which can now be obtained is that

$$v_0^\infty(x,t) := \frac{G(r-t,\theta)}{r}, \quad x = r\theta \tag{8.35}$$

defines an **asymptotic wave function** for $v_0(.,t)$ in $H(\mathbf{R}^3)$ in the sense that

$$\lim_{t\to\infty} \| v_0(.,t) - v_0^\infty(.,t) \| = 0 \tag{8.36}$$

where $\| \cdot \|$ denotes the norm in $H(\mathbf{R}^3)$.

We also notice that (8.33) can be written more compactly in the form

$$G(\tau,\theta) = \Theta h_0(\tau,\theta).$$

The correspondence $\Theta : H(\mathbf{R}^3) \to H(\mathbf{R}\times S^2)$ can be shown to define a unitary operator.

Proofs of these various statements and also of their extensions to more general forms of $\hat{h}_0$ rely on the following convergence results.

Theorem 8.3

Let $\Omega \subset \mathbf{R}^3$ denote an unbounded domain and let $u(x,t)$ have the properties

(i) $u(.,t) \in L_2(\Omega)$ for every $t > t_0$

(ii) $\lim_{t\to\infty} \| u(.,t) \|_H = 0$

where $H := L_2(K \cap \Omega)$ for every compact $K \subset \mathbf{R}^3$

(iii) $| u(x,t) | \leq M / | x |^2$ for every $| x | > r_0$

where t_0, r_0 and M are constants.

Then

$$\lim_{t\to\infty} \| u(.,t) \|_{L_2(\Omega)} = 0.$$

In this section we of course only need the case $\Omega = \mathbf{R}^3$.

Since the real part of the asymptotic wave function v_0^∞ given in (8.35) is also a function of the same form then we can obtain similar results for u_0 as for v_0. The results are summarised in the following Theorem.

Theorem 8.4

Let $\phi_0 \in H(\mathbf{R}^3)$ and $\phi_1 \in D(A_0^{\frac{1}{2}})$ be real valued functions. Let $u_0(x,t)$, defined by means of (8.7) be the solution of the FP. If the asymptotic wave function u_0^∞ is defined by

$$u_0^\infty(x,t) := \frac{F(r-t,\theta)}{r}, \quad x = r\theta \tag{8.37}$$

where $F(\tau,\theta) = \mathrm{Re}(G(\tau,\theta))$ and G is defined by (8.33) with $h_0 = \phi_0 + iA_0^{-\frac{1}{2}}\phi_1 \in H(\mathbf{R}^3)$ then

$$\lim_{t\to\infty} \| u_0(.,t) - u_0^\infty(.,t) \|_{H(\mathbf{R}^3)} = 0. \tag{8.38}$$

Finally in this section we remark that the form of the spectral family $\{E_0(\lambda)\}$ for A_0, defined by (8.23), implies that A_0 is spectrally absolutely continuous (see Theorem 4.5). Indeed, from (8.23) we obtain

$$\| E_0(\mu) f \|^2 = \int_{|p| \leq \sqrt{\mu}} | \hat{f}(p) |^2 \, dp \tag{8.39}$$

for all $\mu \geq 0$ and $f \in H(\mathbf{R}^3)$. Furthermore

$$\frac{d}{d\mu} \| E_0(\mu)f \|^2 = \frac{\mu}{2} \int_{S^2} | \hat{f}(\sqrt{\mu}\eta) |^2 \, d\eta, \quad \mu > 0$$

and for suitable f this quantity is positive for any $\mu > 0$. Consequently, $\sigma(A_0) = \sigma_{ac}(A_0) = \overline{\mathbf{R}}_+$.

8.3 A perturbed problem

For the Perturbed Problem (PP) we take $\Omega \subset \mathbf{R}^3$ to be the unbounded region lying exterior to a closed, smooth, boundary surface $\partial\Omega$. On $\partial\Omega$ we will require that solutions of the wave equation (8.1) should satisfy a homogeneous Neumann condition.

The analysis of the PP is similar in many respects to that of the FP. We introduce here the operator

$$\begin{aligned} A &: L_2(\Omega) \to L_2(\Omega) =: H(\Omega) \\ Au &= -\Delta u, \ u \in D(A) \\ D(A) &= \{u \in H(\Omega) : \ \Delta u \in H(\Omega), \ \partial u/\partial n = 0 \text{ on } \partial\Omega\}. \end{aligned} \tag{8.40}$$

Corresponding to Theorem 8.1 the following result can be obtained.

Theorem 8.5

(i) A is a self-adjoint, non-negative operator on $H(\Omega)$

(ii) $A^{\frac{1}{2}}$ exists with

$D(A^{\frac{1}{2}}) = \{u \in H(\Omega) : \ D^\alpha u \in H(\Omega), \ | \alpha | \leq 1\}$

(iii) $\| A^{\frac{1}{2}} u \|^2 = \sum_{j=1}^{3} \| \partial u/\partial x_j \|^2$ for all $u \in D(A^{\frac{1}{2}})$.

We now interpret (8.1) as

$$\left(\frac{d^2}{dt^2} + A\right) u(t) = 0, \quad t \in \mathbf{R} \tag{8.41}$$

where $u(x,t)$ the unknown in the PP is represented by $u(t)$ and understood to be an $H(\Omega)$ valued function of $t \in \mathbf{R}$. The initial values corresponding to (8.2) will be taken in the form

$$u(0) = \psi_0, \quad u_t(0) = \psi_1 \tag{8.42}$$

where it is assumed that $\psi_0, \psi_1 \in H(\Omega)$.

A solution of the IVP (8.41), (8.42) can be written in the form

$$u(t) = (\cos tA^{\frac{1}{2}})\psi_0 + A^{-\frac{1}{2}}(\sin tA^{\frac{1}{2}})\psi_1. \tag{8.43}$$

If further ψ_0, ψ_1 are real valued functions such that $\psi_0 \in H(\Omega)$ and $\psi_1 \in D(A^{\frac{1}{2}})$ and if we define

$$h := \psi_0 + iA^{-\frac{1}{2}}\psi_1 \tag{8.44}$$

then the solution (8.43) can be represented in the form

$$u(t) \equiv u(.,,t) = \text{Re}\{v(.,t)\} \tag{8.45}$$

where

$$v(t) \equiv v(.,t) = \exp(-itA^{\frac{1}{2}})h \tag{8.46}$$

is the complex-valued solution in $H(\Omega)$ of (8.41), (8.42).

The properties of the operator A stated in Theorem 8.5 indicate that a Spectral Theorem is available for use in interpreting the various coefficients in (8.43) and (8.45). What we now need is a spectral expansion for A which corresponds to that which we obtained for A_0, and as a consequence obtain results similar to (8.18) to (8.20) for A. With this in mind we recall that A_0 has a generalised eigenfunction $w_0(x,p)$ given by (8.16). This eigenfunction is associated with the spectral parameter $|\ p\ |^2$ and corresponds to a plane wave propagating in the direction of the vector p. We would expect therefore, when an obstacle is immersed in the incident field of the FP, that the associated operator A has generalised eigenfunctions which are perturbations of the plane waves arising in a study of the FP. It turns out that there are two families of generalised eigenfunctions for A, which we shall denote by $w_\pm(x,p)$. Mathematically the $w_\pm(x,p)$ must satisfy

$$(\Delta + |\ p\ |^2)w_\pm(x,p) = 0, \quad x \in \Omega \tag{8.47}$$

$$w_\pm(x,p) \in (bc)_0, \quad x \in \partial\Omega. \tag{8.48}$$

Here $(bc)_0$ denotes that homogeneous boundary conditions are required to be satisfied on $\partial\Omega$. In our particular case we will require

$$\frac{\partial w_\pm}{\partial n}(x,p) = 0, \quad x \in \partial\Omega \tag{8.48a}$$

where $\partial/\partial n$ denotes the derivative in the direction of the outward drawn normal to $\partial\Omega$. In constructing the $w_{\pm}$ use will be made of the cu -off function j defined by

$$j(x) = \chi(|x| - r_0), \quad x \in \mathbf{R}^3 \tag{8.49}$$

$\chi \in C(\mathbf{R})$ is such that (i) $\chi(\tau) = 0,\ \tau \leq 0$ (ii) $\chi(\tau) = 1,\ \tau \geq 1$ (iii) $\chi'(\tau) \geq 0,\ \tau \in \mathbf{R}$ and where r_0 is chosen to ensure that $\mathbf{R}^3 \backslash \Omega \subset B(r_0)$; the notation $B(r_0)$ denotes a ball centre the origin and radius r_0.

The eigenfunctions of A are assumed to have the form

$$w_{\pm}(x,p) = j(x)w_0(x,p) + w'_{\pm}(x,p) \tag{8.50}$$

where $w'_{\pm}(x,p)$ represents the **secondary or scattered field** which is produced when an obstacle, represented by $\partial\Omega$, is introduced into the incident field $w_0(x,p)$ discussed in the previous section. For this reason the generalised eigenfunctions $w_{\pm}$ are referred to as **distorted plane** waves.

The determination of the generalised eigenfunctions is completed by requiring that the scattered field w'_{+} should be outgoing whilst w'_{-} should be incoming. Acknowledging this and substituting (8.50) into (8.47)(8.48a) we find that the scattered fields $w'_{\pm}$ should satisfy

$$(\Delta + |p|^2)w'_{\pm}(x,p) = -(\Delta + |p|^2)j(x)w_0(x,p), \quad x \in \Omega \tag{8.51}$$

$$\frac{\partial w'_{\pm}}{\partial n}(x,p) = -\frac{\partial}{\partial n}(j(x)w_0(x,p)), \quad x \in \partial\Omega \tag{8.52}$$

$$\frac{\partial w'_{\pm}}{\partial |x|}(x,p) \mp i|p|w'_{\pm}(x,p) = o(|x|^{-1}), \quad |x| \to \infty \tag{8.53}$$

$$w'_{\pm}(x,p) = 0(|x|^{-1}), \quad |x| \to \infty. \tag{8.54}$$

From a detailed analysis of this boundary value problem the following can be established.

Theorem 8.6

Let $\Omega \subset \mathbf{R}^3$ can be an unbounded region. For every $p \in \mathbf{R}^3$ there exists a unique outgoing (incoming) distorted plane wave $w_+(x,p)(w_-(x,p))$ of the form (8.50) with the property

$$w'_\pm(x,p) = \frac{e^{\pm i|p|r}}{r} T_\pm(\theta,p) + w''_\pm(x,p), \quad x = r\theta \tag{8.55}$$

where $T_\pm \in C^\infty(S^2 \times \{\mathbf{R}^3\backslash 0\})$ and $w''_\pm(x,p) = 0(r^{-2})$ as $r \to \infty$, uniformly for $\theta = x/r \in S^2$ and p in any compact subset of $\mathbf{R}^3\backslash\{0\}$.

The quantities $T_\pm(\theta,p)$ appearing in this Theorem are called the **far-field amplitudes** of the distorted plane waves.

Once the distorted plane waves have been constructed we can obtain the required generalised Fourier transforms. Each of the families $\{w_+(.,p)\}$ and $\{w_-(.,p)\}$ can be shown to define a complete set of generalised eigenfunctions of A. Specifically, it is possible to obtain for A the following analogues of (8.18) and (8.19); for any $f \in H(\Omega)$ the following limits exist

$$(F_\pm f)(p) := \hat{f}_\pm(p) = \lim_{R\to\infty} \int_{\Omega(R)} \overline{w_\pm(x,p)} f(x)dx \tag{8.56}$$

$$f(x) = (F_\pm^{-1} \hat{f}_\pm)(x) = \lim_{R\to\infty} \int_{|p|\leq R} w_\pm(x,p)\hat{f}_\pm(p)dp \tag{8.57}$$

where $\Omega(R) := \Omega \cap \{x \in \mathbf{R}^3 :| x |< R\}$ and where it must be emphasised that the limit in (8.56) is taken in the $H(\mathbf{R}^3)$ whilst that in (8.57) is taken in the $H(\Omega)$ sense. Furthermore it can be shown that the operators

$$F_\pm : H(\Omega) \to H(\mathbf{R}^3) \tag{8.58}$$

are unitary. Hence we shall sometimes write $F_\pm^{-1} = F_\pm^*$.

Therefore when studying the PP we shall use the generalised Fourier transforms defined in (8.56), (8.57) and then argue along similar lines to those followed for the FP to obtain the following result.

Theorem 8.7

For any $f \in H(\Omega)$ and any Lebesgue measurable function Φ

$$\text{(i)} \quad F_\pm \Phi(A) f(p) = \Phi(| p |^2)\hat{f}_\pm(p) \tag{8.59}$$

(ii) $\{E(\lambda)\}$, the spectral family of A has the representation

$$E(\lambda)f(x) = \int_{|p|\leq\sqrt{\lambda}} w_{\pm}(x,p)\hat{f}_{\pm}(p)dp, \quad \lambda \geq 0 \tag{8.60}$$

whence A is a spectrally absolutely continuous operator with spectrum the interval $[0,\infty)$.

Thus we see that F_+ and F_- define spectral representations for A. Consequently they provide a generalisation of the Plancherel theory of Fourier transforms, used for the FP, to exterior domains $\Omega \subset \mathbf{R}^3$. The extension of the results stated here to more general situations is discussed in the references cited in the Commentary to this Chapter.

Now that we have available suitable spectral decompositions for acoustic problems in the exterior region $\Omega \subset \mathbf{R}$ we can proceed along similar lines to those used for the FP but using the generalised Fourier transforms defined above. First we notice that the wave function defined in (8.46) has **two** spectral representations

$$v(x,t) = \lim_{R\to\infty}\int_{|p|\leq R} w_{\pm}(x,p)\exp(-it\mid p\mid)\hat{h}_{\pm}(p)dp \tag{8.61}$$

depending on whether w_+ or w_- is used. These will be called the **outgoing** and **incoming representations** respectively. We would emphasise that the limit in (8.61) must be taken in the $H(\Omega)$ sense. However from (8.56) we notice

$$\hat{h}_{\pm}(p) = \lim_{R\to\infty}\int_{\Omega(R)} \overline{w_{\pm}(x,p)}h(x)dx = F_{\pm}h(p)$$

where the limit is taken in the $H(\mathbf{R}^3)$ sense.

From the representations (8.61) and the results obtained for the FP in §8.2 we can obtain the asymptotic behaviour of the wave function $v(x,t)$ as $t\to\infty$. In principle it does not really matter which of the representations offered by (8.61) is used. It turns out that using the incoming representation is particularly convenient. As in §8.2 we shall only discuss the simplest case here; more general situations will be mentioned in the Commentary to this Chapter. Consequently, we shall consider a complex valued initial state, h, defined as in (8.44) which is such that $h \in H(\Omega)$ and $\hat{h}_{\pm} \in C_0^{\infty}(\mathbf{R}^3)$. These states are dense in $H(\Omega)$ because $C_0^{\infty}(\mathbf{R}^3)$ is dense in $H(\mathbf{R}^3)$ and the operators $F_{\pm} : H(\Omega) \to H(\mathbf{R}^3)$ are unitary.

The wave function corresponding to $\hat{h}_-$ is obtained from (8.61) in the form

$$v(x,t) = \lim_{R\to\infty} \int_{|p|\leq r} w_-(x,p)\exp(-it \mid p \mid)\hat{h}_-(p)dp. \qquad (8.62)$$

To determine the asymptotic behaviour of $v(x,t)$ in (8.62) we recall (8.50) and (8.55) and rewrite (8.62) in the form

$$v(x,t) = v_0(x,t) + v_1(x,t) + v_2(x,t) \qquad (8.63)$$

where

$$v_0(x,t) = \lim_{R\to\infty} \int_{|p|\leq R} w_0(x,p)\exp(-it \mid p \mid)\hat{h}_-(p)dp$$

$$v_1(x,t) = \lim_{R\to\infty} \frac{1}{r} \int_{|p|\leq R} T_-(\theta,p)\exp(-i(r+t) \mid p \mid)\hat{h}_-(p)dp$$

$$v_2(x,t) = \lim_{R\to\infty} \int_{|p|\leq R} w''(x,p)\exp(-it) \mid p \mid)\hat{h}_-(p)dp$$

where the limits are taken in the $H(\Omega)$ sense.

If we recall (8.10) and the spectral decompositions for the FP indicated in (8.18) to (8.20) then we see that $v_0(x,t)$ can also be written meaningfully in the form

$$v_0(x,t) = \exp(-itA_0^{\frac{1}{2}})g_0(x) \qquad (8.64)$$

where $v_0(x,0) = g_0(x)$. Furthermore, using (8.56), (8.57)

$$g_0 = F^*\hat{h}_- = F^*F_-h \in H(\mathbf{R}^3) \qquad (8.65)$$

the inclusion sign follows since $F : H(\mathbf{R}^3) \to H(\mathbf{R}^3)$ and $F_\pm : H(\Omega) \to H(\mathbf{R}^3)$. Thus we see that $v_0(x,t)$ solves a FP in $H(\mathbf{R}^3)$; that is, it represents the wave field in the absence of a scattering target. The principal result which can now be obtained is that $v(x,t)$ is asymptotically equal to this wave as $t \to \infty$. For more general initial conditions than those we have been considering here the denseness of $C_0^\infty(\mathbf{R}^3)$ in $H(\mathbf{R}^3)$ is used and the following result can be obtained.

Theorem 8.8

Let $\Omega \subset \mathbf{R}^3$ be an exterior region, then for every $h \in H(\Omega)$

(i) $\lim_{t\to\infty}(v_1(.,t)+v_2(.,t))=0.$ (8.66)

where the limit is taken in the $H(\Omega)$-sense.

(ii) $\lim_{t\to\infty}\| v(.,t)-v_0(.,t)\|_{H(\Omega)}=0.$ (8.67)

If we now recall the analysis of §8.2 then we see that the function v_0 defined in $H(\mathbf{R}^3)$ by

$$v_0(.,t)=\exp(-itA_0^{\frac{1}{2}})h_0$$

where $h_0 := F^*F_-h$, with h the initial value for the PP is, in fact, a wave function for the FP in $H(\mathbf{R}^3)$. The results of §8.2 show that this wave function has an associated asymptotic wave function $v_0^\infty(.,t)$ defined by (8.35) with the property

$$\lim_{t\to\infty}\| v_0(.,t)-v_0^\infty(.,t)\|_{H(\mathbf{R}^3)}=0. \tag{8.68}$$

As in (8.35)

$$v_0^\infty(x,t)=\frac{G(r-t,\theta)}{r},\quad x=r\theta \tag{8.69}$$

but in this particular case we have, acknowledging (8.65),

$$G=\Theta h_0=\Theta F^*F_-h. \tag{8.70}$$

We notice that the only difference between the asymptotic wave function (8.35) for the FP and the asymptotic wave function (8.69) for the PP is that in the former the Fourier transform F is used whilst in the latter the generalised Fourier transform F_- is used. Consequently, $\hat{h}_0 = Fh_0$ is replaced by $\hat{h}_- = F_-h$.

A more general form of these results can be stated as follows.

Theorem 8.9

For any $h\in H(\Omega)$ the wave function $v_0^\infty(.,t)$ defined in (8.69), (8.70) is an asymptotic wave function in $H(\Omega)$ for the wave $v(x,t)$ defined in (8.46); that is

$$\lim_{t\to\infty}\| v(.,t)-v_0^\infty(.,t)\|_{H(\Omega)}=0. \tag{8.71}$$

The profile of the asymptotic wave function v_0^∞ in (8.69) is defined by the function G given in the form (see (8.33))

$$G(\tau,\theta)=\frac{1}{(2\pi)^{\frac{1}{2}}}\int_0^\infty e^{i\tau\rho}\hat{h}_-(\rho\theta)(-i\rho)d\rho. \tag{8.72}$$

8.4 Concerning the wave operators

We have shown in §8.2 that the FP has a complex-valued solution defined by

$$v_0(x,t) = \exp(-itA_0^{\frac{1}{2}})h_0(x) =: U_0(t)h_0(x) \tag{8.73}$$

and in §8.3 that the PP has a complex-valued solution defined by

$$v(x,t) = \exp(-itA^{\frac{1}{2}})h(x) =: U(t)h(x). \tag{8.74}$$

We have also seen that there is an asymptotic wave function v_0^∞ such that

$$\lim_{t\to\infty} \| v_0(.,t) - v_0^\infty(.,t) \|_{H(\mathbf{R}^3)} = 0 \tag{8.36}$$

$$v_0^\infty(x,t) = \frac{G(r-t,\theta)}{r}. \tag{8.35}$$

Furthermore we have seen that

$$\lim_{t\to 0} \| v(.,t) - v_0(.,t) \|_{H(\Omega)} = 0 \tag{8.67}$$

where $v_0(x,t) = U_0(t)g_0$ with

$$g_0 = F^* F_- h. \tag{8.65}$$

We have also seen that

$$\lim_{t\to\infty} \| v(.,t) - v_0^\infty(.,t) \|_{H(\Omega)} = 0 \tag{8.71}$$

with the associated profile of the asymptotic wave function defined as in (8.72). (See also the remarks following (8.70.))

In this section we want to compare the wave fields represented by v_0 and v in a more direct manner. Our aim has always been to determine whether or not the solutions of the FP and PP are asymptotically equal. Clearly this will be the case when

$$\lim_{t\to\pm\infty} \| v(.,t) - v_0(.,t) \|_{H(\Omega)} = 0. \tag{8.75}$$

If we introduce the mapping

$$J(\Omega) : H(\Omega) \to H(\mathbf{R}^3) \tag{8.76}$$

$$J(\Omega)v(x) = \begin{cases} v(x), & x \in \Omega \\ 0, & x \in \mathbf{R}^3 \backslash \Omega \end{cases}$$

then acknowledging (8.73) and (8.74) we see that (8.75) is equivalent to

$$\lim_{t \to \pm\infty} \| J(\Omega)U(t)h - U_0(t)h_0 \|_{H(\mathbf{R}^3)} = 0. \tag{8.77}$$

Since U_0 defined in (8.73) is unitary on $H(\mathbf{R}^3)$ then we can write (8.77) in the form

$$\begin{aligned} 0 &= \lim_{t \to \pm\infty} \| U_0^{-1}(t)J(\Omega)U(t)h - h_0 \|_{H(\mathbf{R}^3)} \\ &= \lim_{t \to \pm\infty} \| W(t)h - h_0 \|_{H(\mathbf{R}^3)} \\ &= \| W_\pm h - h_0 \|_{H(\mathbf{R}^3)} \end{aligned} \tag{8.78}$$

where we have defined

$$W_\pm := \lim_{t \to \pm\infty} W(t) = \lim_{t \to \pm\infty} U_0^*(t)J(\Omega)U(t). \tag{8.79}$$

The quantities $W_\pm$ are the **wave operators** associated with $A_0^{\frac{1}{2}}$, $A^{\frac{1}{2}}$ and $J(\Omega)$.

Thus we see from (8.75) and (8.78) that the FP and the PP are asymptotically equal as $t \to \pm\infty$ provided the initial data for the two problems are related according to

$$W_\pm h = h_0^\pm \tag{8.80}$$

where the superscripts on h_0 indicate the possibility of having to use different initial values for the FP when $t \to +\infty$ and when $t \to -\infty$.

To prove the existence of the limits in (8.79), which define the wave operators $W_\pm$, is a considerable undertaking as indeed is the proof of the completeness of the wave operators. Results which have been obtained in this connection for the wave equation can be summarised in the following theorems.

Theorem 8.10

Let $\Omega \subset \mathbf{R}^3$ be an unbounded region and define

$$W_+ = W_+(A_0^{\frac{1}{2}}, A^{\frac{1}{2}}, J(\Omega)) = s- \lim_{t \to \infty} U_0^*(t)J(\Omega)U(t) \tag{8.81}$$

then the wave operator

$$W_+ : H(\Omega) \to H(\mathbf{R}^3)$$

exists, is unitary and

$$W_+ = F^* F_-. \tag{8.82}$$

Furthermore, it defines a unitary equivalence between $\{E(\lambda)\}$, the spectral family of the operator A in the PP, and $\{E_0(\lambda)\}$ the spectral family of the operator A_0 in the FP according to

$$E(\lambda) = W_+^* E_0(\lambda) W_+ \quad \text{for all } \lambda \in \mathbf{R}. \tag{8.83}$$

Theorem 8.11

Let $\Omega \subset \mathbf{R}^3$ be an exterior region and define

$$W_- := W_-(A_0^{\frac{1}{2}} A^{\frac{1}{2}}, J(\Omega)) = s-\lim_{t \to -\infty} U_0^*(t) J(\Omega) U(t) \tag{8.84}$$

then the wave operator

$$W_- : H(\Omega) \to H(\mathbf{R}^3)$$

exists, is unitary and

$$W_- = F^* F_+. \tag{8.85}$$

Furthermore, the spectral families $\{E(\lambda)\}$ and $\{E_0(\lambda)\}$ introduced in Theorem 8.11 are unitarily equivalent in the form

$$E(\lambda) = W_-^* E_0(\lambda) W_- \quad \text{for all } \lambda \in \mathbf{R} \tag{8.86}$$

8.5 Summary and additional comments

In this Chapter we have considered two time-dependent problems. A FP represented by (8.1)-(8.3) with $\Omega = \mathbf{R}^3$ and a PP represented by (8.1)-(8.3) with $\Omega \subset \mathbf{R}^3$ an unbounded region with a smooth boundary $\partial\Omega$ on which were imposed Neumann conditions. We have seen that the FP has a solution which can be expressed as a complex-valued wave function of the form

$$v_0(x,t) = \exp(-itA_0^{\frac{1}{2}}) h_0(x). \tag{8.10}$$

The operator A_0 can be shown to be self-adjoint and non-negative on $H(\mathbf{R}^3)$ and consequently the coefficients in (8.10) can be interpreted by means of the Spectral Theorem. However, for the wave equation in $\mathbf{R}^3$ a more immediate way of obtaining a solution is to use the usual Fourier transforms on $\mathbf{R}^3$ but written in the form (8.18)-(8.20). These forms exhibit the connection with $\sigma(A_0)$ and as such provide an expansion theorem in terms of w_0, the generalised eigenfunctions of A_0. The expansion theorem enabled us to interpret the coefficients in (8.10) without having to invoke the Spectral Theorem directly. However, we could display, in (8.23), a result which provided the connection between expansions in terms of generalised eigenfunctions of A_0 and the spectral family of A_0 required by the Spectral Theorem. Finally, we discussed the asymptotic behaviour of solutions to the FP obtained in terms of these generalised eigenfunctions and introduced such concepts as wave function, asymptotic wave function and wave profile. Questions of existence and uniqueness of solution were settled in Chapter 7 and here the main concern was to indicate a constructive method of solution. Specifically, (8.7) and (8.18)-(8.20) can be combined to yield a representation of the solution to the FP in the form

$$u_0(x,t) = \int_{\mathbf{R}^3} w_0(x,p)\{\hat{\phi}_0(p)\cos t \mid p \mid + \hat{\phi}_1(p)\frac{\sin t \mid p \mid}{\mid p \mid}\}dp. \tag{8.87}$$

In developing this approach we notice that although we have been principally concerned with a time-dependent problem nevertheless we have had to discuss an associated time-independent, that is stationary, problem involving A_0 and its spectral properties.

This is a successful strategy for the FP and the intention has been to develop a similar one for the PP. The analysis of the PP can proceed along very similar lines to those adopted for the FP. Indeed, the solution of the PP can be expressed as a complex-valued wave function in the form

$$v(x,t) = \exp(-itA^{\frac{1}{2}})h(x). \tag{8.46}$$

To interpret this solution form a symbolically very similar expansion theorem to that obtained for the FP was obtained in (8.56), (8.57). However, we notice that now either one or other of the distorted plane waves $w_\pm$ had to be used rather than the free space plane wave. Nevertheless, similar notions of wave function, asymptotic wave function and wave profile could be introduced and constructed in terms of these generalised Fourier transforms. Once again we noticed that although we were dealing with a time-dependent problem there is an associated stationary problem centred on the operator A and its spectrum $\sigma(A)$. Results which can be obtained in this connection are summarised in the following theorems.

Theorem 8.12

The self-adjoint operator A corresponding to the PP has no eigenvalues.

Theorem 8.13

The subspaces $H_{ac}(A)$ and $H_{sc}(A)$ are subspaces of $H(\Omega)$ which are orthogonal and reduce the operator A. Furthermore, $H(\Omega)$ can be decomposed in the form $H(\Omega) = H_{ac}(A) \oplus H_{sc}(A)$.

Theorem 8.14

For $\Omega \subset \mathbf{R}^3$ an unbounded region

$$H(\Omega) = H_{ac}(A).$$

The proof of this last theorem, which is so important when developing a scattering theory, is based on Stones Theorem and the Limiting Absorption Principle (LAB).

The LAB provides a method for determing the existence of solutions to the stationary problem. It is based on noticing that if A is a self-adjoint operator on $H(\Omega)$ and $\lambda = \mu + i\nu \in \mathbf{C}$ with $\nu \neq 0$ then the equation

$$(A - \lambda I)u(x, \lambda) = f(x)$$

has a solution $u(., \lambda) \in H(\Omega)$ for each $f \in H(\Omega)$ because $\lambda \notin \sigma(A)$. In the LAB method we look for solutions of

$$(A - \lambda I)u(x, \mu) = f \tag{8.88}$$

in the form

$$u_{\pm}(x, \mu) = \lim_{\nu \to 0\pm} u(x, \lambda). \tag{8.89}$$

The difficulty with this approach, appealing as it is, is centred on the interpretation of the limit in (8.89). Ideally, for our purposes, this limit should be understood, if possible, in an $H(\Omega)$ sense as this would then automatically provide uniform convergence with respect to x. Since, in general, solutions of (8.88) do not have solutions in $H(\Omega)$ [20, 52, 79] the limit in (8.89) can only be understood in the sense of convergence in $H(\omega)$ where ω is an arbitrary, bounded subdomain of Ω.

A justification of the LAB requires a proof of the existence of the limit in (8.89). Standard theory of elliptic equations [86] will then indicate that $u_{\pm}$ solve (8.88).

Many of the difficulties associated with having to work in unbounded, sub domains of Ω with application of the Poincaré inequality and Sobolev embedding theorems can be eased by working in a weighted Sobolev space structure [54, 63, 64, 65, 66, 67].

Once the validity of the limit in (8.89) has been proved then the existence of the solution will have been settled.

For the particular case of the distorted plane waves $w_{\pm}$ we see that their uniqueness is established by requiring that the scattered or secondary field should satisfy a Sommerfeld radiation condition whilst their existence is settled by proving the validity of LAB.

An important result, which has been displayed in Theorem 8.9 is that every solution of the PP in $H(\Omega)$ is asymptotically equal to a solution of the FP in $H(\mathbf{R}^3)$. As a consequence of this we were able to indicate, in Theorem 8.10 and Theorem 8.11, that the associated wave operators exist and also provide an explicit representation for them.

Chapter 9
Nonlinear scattering theory

9.1 Introduction

Nonlinear problems tend to be dealt with on an individual basis since the governing equations frequently have special features which exert a particularly dominating influence on the solutions. Nevertheless, the basic requirements are the same for all problems be they linear or nonlinear; we need information about the existence and uniqueness of solution. Once this is available we can then investigate other aspects of the solution such as their regularity and their asymptotic behaviour; in particular we can develop scattering theories. In the literature these various aspects are often treated separately for each particular problem. However, it turns out that many nonlinear problems in the applied sciences present certain common problems in abstract functional analysis. Indeed using standard linear functional analysis together with the Contraction Mapping Principle considerable progress can be made in the study of these nonlinear problems. The abstract approach can offer a unified approach and provide a means of clarifying those properties of the solution which are general and those which are dependent on the special features of the equation being considered.

To illustrate an abstract approach consider the following IVP; determine a quantity $u(x,t)$ satisfying

$$\left(\frac{\partial^2}{\partial t^2} - \Delta\right) u(x,t) = -f(u), \quad (x,t) \in \mathbf{R}^n \times \mathbf{R} \tag{9.1}$$

$$u(x,0) = u_0(x), \quad u_t(x,0) = u_1(x) \tag{9.2}$$

where f is a given, nonlinear function of u. Typical examples are

Nonlinear Klein-Gordon Equation (NLGK)

$$f(u) = (m^2 - a \mid u \mid^{p-1})u =: (m^2 - g(\mid u \mid))u \tag{9.3}$$

Nonlinear Wave Equation (NLW)

$$f(u) = u^p. \tag{9.4}$$

A discussion of the IVP (9.1), (9.2) can be given an abstract setting by first reducing it to a first order system using the, now familiar, technique outlined in the Introduction and used in subsequent chapters. Specifically, if we set $u_t(x,t) = v(x,t)$ then the IVP with f of the form (9.3) can be reformulated as

$$\begin{bmatrix} u \\ v \end{bmatrix}_t (x,t) - \begin{bmatrix} 0 & I \\ \Delta - m^2 & 0 \end{bmatrix} \begin{bmatrix} u \\ v \end{bmatrix} (x,t) = \begin{bmatrix} 0 \\ g(|u|) \end{bmatrix} (x,t)$$

$$\begin{bmatrix} u \\ v \end{bmatrix} (x,0) = \begin{bmatrix} u_0 \\ u_1 \end{bmatrix}.$$

This can be written more compactly, with an obvious notation, in the form

$$\phi'(x,t) - M\phi(x,t) = N(\phi)(x,t) \tag{9.5}$$

$$\phi(x,0) = \phi_0(x). \tag{9.6}$$

The differential expression $(-\Delta + m^2)$ generates an operator A defined by

$$A = L_2(\mathbf{R}^n) \to L_2(\mathbf{R}^n) =: H(\mathbf{R})^n \tag{9.7}$$

$$Au = (-\Delta + m^2)u \quad \text{for all } u \in D(A)$$

$$D(A) = \{u \in H(\mathbf{R}^n) : (-\Delta + m^2)u \in H(\mathbf{R}^n)\}.$$

We remark that $D(A)$ can also be characterised by requiring that $(k^2 + m^2)\hat{f}(k) \in H(\mathbf{R}^n)$ where $\hat{f}(k)$ denotes the usual Fourier transform of $f(x)$.

The following properties of A can be established as in previous sections.

(i) A is non-negative, self-adjoint on $H(\mathbf{R}^n)$

(ii) A has a unique square root

(iii) $A^{\frac{1}{2}}$, the positive square root, is a closed operator.

From (iii) it follows that $D(A^{\frac{1}{2}})$ is a Hilbert space with respect to the inner product $(A^{\frac{1}{2}}u, A^{\frac{1}{2}}v)_{H(\mathbf{R}^n)}$. Using this fact we now introduce the Hilbert space $h := D(A^{\frac{1}{2}}) \oplus H(\mathbf{R}^n)$ with inner product

$$((u_1, v_1), (u_2, v_2)) = (A^{\frac{1}{2}}u_1, A^{\frac{1}{2}}u_2)_{H(\mathbf{R}^n)} + (v_1, v_2)_{H(\mathbf{R}^n)}. \tag{9.8}$$

This space is, as the notation suggests, equivalent to the energy space, H_E, introduced in earlier chapters.

Now define

$$
\begin{aligned}
&G : H \to H \\
&G\phi = i\begin{bmatrix} 0 & I \\ -A & 0 \end{bmatrix}\begin{bmatrix} u \\ v \end{bmatrix} = -iM\phi, \quad \phi \in D(G) \\
&D(G) = \{\phi \in H : M\phi \in H.\}
\end{aligned} \tag{9.9}
$$

Then proceeding as in previous chapters we can show that G is symmetric and is closed since A and $A^{\frac{1}{2}}$ are closed.

The IVP (9.5), (9.6) can now be reformulated as a problem in H: determine $\phi(.,t) = \phi(t) \in H$ which satisfies

$$\phi'(t) + G\phi(t) = N(\phi)(t) \tag{9.10}$$

$$\phi(0) = \phi_0 \tag{9.11}$$

where, we emphasise, ϕ is now regarded as an H-valued function of t.

Thus we see that the IVP (9.1)(9.2) is a special case of an abstract class of Hilbert space problems. Namely, given a self-adjoint operator G on a Hilbert space H, an element $\phi \in H$ and a nonlinear mapping, N, of H into itself determine ϕ, an H-valued function of t, which satisfies (9.10), (9.11).

The equation (9.10) can be integrated, just as for the linear case ($N \equiv 0$) however in this case rather than obtaining an actual representation of the required solution we now obtain an integral equation, equivalent to (9.10), (9.11), of the form

$$\phi(t) = U_0(t)\phi_0 + \int_0^t U_0(t-s)N(u(s))ds \tag{9.12}$$

where $U_0(t) = \exp(-itG)$. We see that when $N \equiv 0$ then we obtain the solution of the related linear equation; and it is in the form discussed in previous chapters. Thus we might expect that much of the analysis leading to a scattering theory for the linear problem will appear in the corresponding analysis for the non-linear problem (9.12). Indeed when developing an associated scattering theory we take the case $N \equiv 0$ to define the Free Problem (FP) whilst the Perturbed Problem (PP) will arise in the case when $N \not\equiv 0$.

9.2 Concerning existence of solutions

A term of the form ϕ^p, $p > 1$ in an equation such as (9.10) will tend to magnify the solution when $|\phi|$ is large, that is, to cause the solution to **blow up**, but will tend to have negligible effect when $|\phi|$ is small. On the other hand solutions of a FP associated with (9.10) with nice initial data, are known to decay in the supremum norm, as $t \to \infty$ like $t^{-n/2}$ for the Klein-Gordon equation and $t^{-(n-1)/2}$ for the wave equation. This raises the possibility that there are circumstances under which (9.10), the PP, will have solutions having the same rate of decay. Consequently, the possibility of being able to develop a scattering theory would seem to depend on a subtle interplay between rates of decay and the degree of the nonlinearity present. The basic idea is that if the initial data is small and the degree of nonlinearity sufficiently large then the nonlinear terms should remain small in comparison with the linear terms. Alternatively, we could say that small initial data should be capable of preventing blow up of the solution for a time which is long enough so that the nonlinear term gets "turned off" as $t \to \infty$ and the linear terms become dominant.

These several remarks indicate that in order to be able to develop a scattering theory in a non-linear setting then we must be assured of the global existence of solutions to the problems; that is the existence of solutions which exist for all time.

To establish the global existence of solutions to the IVP (9.10), (9.11) we first reformulate it as the integral equation (9.12) and then solve this equation by the Contraction Mapping Principle. This leads to the following local existence theorem.

Theorem 9.1

let G be a self-adjoint operator on a Hilbert space H and N a, not necessarily linear, mapping from $D(G)$ into $D(G)$ with the properties

(i) $\| N(\phi) \| \leq C(\| \phi \|) \| \phi \|$

(ii) $\| GN(\phi) \| \leq C(\| \phi \|, \| G\phi \|) \| G\phi \|$

(iii) $\| N(\phi_1) - N(\phi_2) \| \leq C(\| \phi_1 \|, \| \phi_2 \|) \| \phi_1 - \phi_2 \|$

(iv) $\| G(N(\phi_1) - N(\phi_2)) \| \leq C(\| \phi_1 \|, \| G\phi_1 \|, \| \phi_2 \|, \| G\phi_2 \|) \| G\phi_1 - G\phi_2 \|$

for all $\phi, \phi_1, \phi_2 \in D(G)$ where each constant C is a monotone increasing, bounded function of the indicated norms. Then, for each $\phi_0 \in D(G)$ there is a T, depending on $\| \phi_0 \|$ and $\| G\phi_0 \|$ such that (9.10), (9.11) has a unique, continuously differentiable solution for all $t \in [0, T]$.

Furthermore, for each subset of H of the form

$$\{\phi \in H : \| \phi \| < a, \| G\phi \| < b\}$$

the quantity T can be chosen uniformly for all ϕ_0 in this subset.

Armed with such a local existence result as Theorem 9.1 it is often possible to obtain a global result; typically the following holds.

Theorem 9.2

Let G, N and H be as in Theorem 9.1 with hypothesis (ii) replaced by

(ii)′ $\| GN(\phi) \| \leq C(\| \phi \|) \| G\phi \|$.

If on every finite interval for which a strong solution of (9.10), (9.11) exists $\| \phi(T) \|$ is bounded then (9.10), (9.11) has a unique global solution.

As an illustration of how a local result can lead to a global result suppose we have a local existence theorem with an existence time T which depends only on $\| \phi_0 \|$, the norm of the initial data. For emphasis write $T = T(\| \phi_0 \|)$. We now solve the IVP, with initial date ϕ_0, up to the time $T_1 := T(\| \phi_0 \|)$. We repeat the process, with initial data $\phi(T_1)$, up to the time $T_2 = T(\| \phi(T_1) \|)$ and so on. Consequently, we have existence in the intervals defined by $0 < T_1 < T_2 < \ldots$. If now we show that the T_m converge to some limit $T^* \leq \infty$ then we have existence on $[0, T^*)$. If further we know a priori that $\| \phi(t) \|$ is bounded then the sizes of the above intervals does not tend to zero and we have global existence.

Ideally, for our purposes when developing a scattering theory we want $T^* = \infty$. To ensure this it turns out that some control is needed over the asymptotic behaviour of solutions. For instance, if typically we assume in (9.1) that $f(u) = 0(| u |^p)$ as $u \to 0$ then we will need the degree p to be high enough to ensure that $f(u)$ is small enough. We know that for the linear equation the fastest rate of decay of the solution as $t \to \infty$ is $0(t^{-q})$ where $q = n/2$ for the Klein-Gordon equation and $q = (n-1)/2$ for the wave equation. Thus, plausible conditions for the global existence of small amplitude solutions could be

(i) $p > 1 + 2/n$, NLKG

(ii) $p > 1 + 2/(n-1)$, NLW.

9.3 Scattering theory

As in the previous section we shall concentrate on the IVP (9.10), (9.11) and its equivalent integral equation form (9.12). We take the problem (9.10), (9.11) as the PP and the same IVP with $N \equiv 0$ as the FP. Solutions of the PP are called **scattering states** and those of the FP are called **free states**. A basic property of a scattering state is that interaction effects, be they due to either a target or some potential-type term, become negligible in the distant past or far future. The scattering state thus looks like a free state as $t \to -\infty$ and like some other free state as $t \to +\infty$. Since the PP is taken as the given non-linear problem and the FP the associated FP then this situation can only be obtained if the non-linear term, the perturbation, is switched off as $t \to \pm\infty$.

Just as for the linear problems discussed earlier when we come to develop a scattering theory for (9.12) we would like to prove that there exists a subset $H_s \subset H$, the scattering subset for (9.12), which has the properties

(i) For each $\phi_- \in H_s$ there is a solution ϕ of (9.12) which, as $t \to -\infty$, tends to the solution of the FP with initial data ϕ_-, that is $\phi(t) \to U_0(t)\phi_-$ as $t \to -\infty$.

(ii) For each $\phi_+ \in H_s$ there is a solution $\tilde{\phi}$ of (9.12) which, as $t \to +\infty$, tends to the solution of the FP with initial data ϕ_+, that is

$$\tilde{\phi}(t) \to U_0(t)\phi_+ \quad \text{as } t \to +\infty.$$

If (i), (ii) can be established then we define wave operators

$$W_- : \phi_- \to \phi(0) \quad \text{and} \quad W_+ : \phi_+ \to \tilde{\phi}(0).$$

Furthermore, if we can prove

(iii) $R(W_+) = R(W_-)$

the so-called **asymptotic completeness of $W_\pm$.**

(iv) $W_\pm$ are one-one operators

then we can define the **scattering operator**

$$S = (W_+)^{-1}W_- : H_s \to H_s$$

and discuss its properties.

In developing a scattering theory for the IVP (9.10), (9.11) with initial data ϕ_- given at $t = -\infty$ we proceed, as in the linear case, by the following steps.

Step 1: Show that (9.10), (9.11) with initial data ϕ_- at $t = -\infty$ is equivalent to the integral equation

$$\phi(t) = U_0(t)\phi_- + \int_{-\infty}^{t} U_0(t-s)N(\phi)(s)ds. \tag{9.13}$$

Step 2: Show that (9.13) has a global solution.

Step 3: Show that $\phi(t) \to U_0(t)\phi_-$ as $t \to -\infty$.

Step 4: Show that there exists a ϕ_+ so that $\phi(t) \to U_0(t)\phi_+$ as $t \to +\infty$.

For nonlinear problems, however, these steps are much more difficult to establish than for linear problems. However, if the initial data ϕ_- at $t = -\infty$ are sufficiently small then **Step 2** can be established. If further the solutions $U_0(t)\phi_\pm$ of the free problems decay sufficiently rapidly and if N has a large enough degree of nonlinearity then the required scattering theory can be developed. A first result in this connection is the following.

Theorem 9.3

Let $\{U_0(t)\}_{t\in\mathbf{R}}$ be a group of linear, isometric operators on H. Let $N : H \supset D(N) \to H$ be a, not necessarily linear, operator and let $\phi(t)$ be a function with values in $D(N)$ such that

(i) $(N\phi)(t)$ is continuous in H

(ii) $$\frac{d}{dt}\{U_0(t)\phi(t)\} = U_0(-t)N(\phi)(t). \tag{9.14}$$

If

$$\int_{-\infty}^{\infty} \| N\phi(t) \| dt < \infty \tag{9.15}$$

then there exist $\phi_\pm \in H$ such that

$$\| \phi(t) - U_0(t)\phi_\pm \| \to 0 \quad \text{as } t \to \pm\infty. \tag{9.16}$$

Proof: Notice first that the vector ϕ_+ defined by

$$\phi_+ = U_0(-t)\phi(t) + \int_t^{+\infty} U_0(-s)N(\phi)(s)ds$$

is, by (9.14) independent of t. Consequently, operating on both sides with $U_0(t)$ and taking norms yields

$$\begin{aligned} \| U_0(t)\phi_+ - \phi(t) \| &\leq \int_t^{\infty} \| U(t-s)N(\phi)(s) \| \, ds \\ &\leq \int_t^{\infty} \| N(\phi)(s) \| \, ds. \end{aligned}$$

Using (9.1) we see that the right hand side of this inequality tends to zero as $t \to +\infty$. A similar result can be obtained when $t \to -\infty$. Hence (9.16) is established. ∎

Integrating (9.10) we can obtain the convenient form

$$\phi(t) = U_0(t-T)\phi(T) + \int_T^t U_0(t-s)N(\phi)(s)ds. \tag{9.17}$$

Now $U_0(t-T)\phi(T) = U_0(t)U_0(-T)\phi(T)$, so if we define

$$\phi_\pm = \lim_{T \to \pm\infty} U_0(-T)\phi(T)$$

then we obtain from (9.17), in the limit as $T \to \pm\infty$,

$$U_0(t)\phi_\pm = \phi(t) + \int_t^{\pm\infty} U_0(t-s)N(\phi)(s)ds. \tag{9.18}$$

Therefore, we see that in arriving at (9.18) we have completed **Steps 1** to **4** provided that Theorem 9.3 is available for our use. To ensure that this is indeed the case we see that it remains to establish that there exists a $\phi(t)$ satisfying (9.14); that is we must establish the global existence of the solution to (9.14). To be able to give the main result in this direction we need some preparation.

Let $G : H \to H$ be a self-adjoint operator on H, a Hilbert space with norm $\| \cdot \|$. Suppose there are two other "norms", denoted $\| \cdot \|_1$ and $\| \cdot \|_2$, which can be defined on H and which satisfy all the properties of a norm save that $\| \phi \|_1 = 0$ need not imply $\phi = 0$ and $\| \cdot \|_2$ might possibly become infinite. We shall assume

A1: There exists $c > 0$ such that $\| \phi \|_1 \leq c \| \phi \|$ for all $\phi \in H$.

A2: There exist constants $c_1 > 0$ and $d > 0$ such that for all $\phi \in H$

$$\| U_0\phi \|_1 \leq c_1 t^{-d} \| \phi \|_2, \quad | t | \geq 1.$$

A3: There exist constants $\beta > 0$, $\delta > 0$ and $q \geq 1$ with $dq > 1$ such that for all $\phi_1, \phi_2 \in H$ satisfying $\| \phi_i \| \leq \delta$, $i = 1, 2$

$$\| N(\phi_1) - N(\phi_2) \| \leq \beta(\| \phi_1 \|_1 + \| \phi_1 \|_1)^q \| \phi_1 - \phi_3 \|$$

$$\| N(\phi_1) - N(\phi_2) \|_2 \leq \beta\{(\| \phi_1 \|_1 + \| \phi_2 \|_1)^{q-1} \| \phi_1 - \phi_2 \|_1 + \\ + (\| \phi_1 \|_1 + \| \phi_2 \|_1)^q \| \phi_1 - \phi_2 \|\}.$$

We now define, for any H-valued element $\psi(t)$, $t \in \mathbf{R}$,

$$||| \psi ||| = \sup_{t\in\mathbf{R}} \| \psi(t) \| + \sup_{t\in\mathbf{R}} (1 + | t |)^d \| \psi(t) \|_1 \tag{9.19}$$

and introduce

$$H_s := \{\phi \in H : ||| U_0(t)\phi ||| < \infty\} \tag{9.20}$$

$$\| \phi \|_s := ||| U_0(t)\phi ||| .$$

Thus we see that the scattering states are simply those elements in H which decay, 'nicely', under free propagation.

We can now state the following global existence result for small data.

Theorem 9.4

Let G be a self-adjoint operator on a Hilbert space H and let N be a, not necessarily linear, operator mapping H into itself. Suppose that there exist norms on H such that **A1** to **A3** hold. Then there exists a constant $\eta > 0$ such that for all $\phi_- \in H_s$ with $\| \phi_- \|_s < \eta$ the equation

$$\phi(t) = U_0(t)\phi_- + \int_{-\infty}^{t} U_0(t - s)N(\phi)(s)ds \tag{9.21}$$

has a global H-valued solution ϕ with $||| \phi ||| \leq 2\eta$. Furthermore

(i) for each t the solution $\phi(t) \in H_s$

(ii) $\| \phi(t) - U_0(t)\phi_- \| \to 0$ as $t \to -\infty$.

A companion result is the following.

Theorem 9.5

Suppose that all the hypotheses of Theorem 9.4 hold and let $\phi(t)$ be a solution of (9.21) corresponding to $\phi_- \in H_s$ with $\| \phi_- \|_s \leq \eta$. For η sufficiently small

(i) there exists a $\phi_+ \in H_s$ with $\| \phi_+ \|_s \leq 2\eta$ such that

$$\| \phi(t) - U_0\phi_+ \| \to 0 \quad \text{as } t \to +\infty$$

(ii) the mapping $S : \phi_- \to \phi_+$ is a one to one and continuous, in the $\| \cdot \|$-topology, mapping of the ball $\{\psi \in H_s : \| \psi \|_s \leq \eta\}$ into the ball $\{\psi \in H_s : \| \psi \|_s \leq 2\eta\}$

(iii) $\| U_0(-t)\phi(t) - \phi_+ \| \to 0$ as $t \to +\infty$

(iv) S is continuous in the $\| \cdot \|_s$-topology.

With these two results available the programme **Step1** to **Step 4** is complete.

Although we have concentrated in the above on an IVP with initial data ϕ_- at $t = -\infty$ nevertheless similar results concerning global existence can be obtained when initial data ϕ_0 is given at $t = 0$ provided $\| \phi_0 \|_s$ is small enough. For this IVP the equivalent integral equation has the form

$$\phi(t) = U_0(t)\phi_0 + \int_0^t U_0(t - s)N(\phi)(s)ds. \tag{9.22}$$

Once local existence results are available we can solve this equation, by iteration for example, and obtain a solution in the form $M(t)\phi_0$, say, where $M(T) : \phi_0 \to \phi(t)$.

A first result concerning the existence of wave operators is given by the following.

Theorem 9.6

Suppose that all the hypotheses of Theorem 9.5 hold for (9.22). Suppose also that for each η and T the solutions $M(t)\phi_0$ of (9.22) are uniformly bounded in the $\| \cdot \|$-norm for all $\| \phi_0 \| \leq \eta$ and for all $0 < | t | \leq T$. Then the following results hold.

(i) For each $\phi_- \in H_s$ there is a global solution $\phi(t)$ of (9.22) such that $\phi(t) \in H_s$ for each t and

$$\| \phi(t) - U_0(t)\phi_- \| \to 0 \quad \text{as } t \to -\infty$$
$$\| U_0(-t)\phi(t) - \phi_- \|_s \to 0 \quad \text{as } t \to -\infty.$$

(ii) The mapping $W_- : \phi_- \to \phi_0$ maps H_s into H_s and is uniformly continuous on balls in H_s.

(iii) For each $\phi_+ \in H_s$ there is a global solution $\phi(t)$ of (9.22) such that $\phi(t) \in H_s$ for each t and

$$\| \phi(t) - U_0(t)\phi_+ \| \to 0 \quad \text{as } t \to +\infty$$
$$\| U_0(-t)\phi(t) - \phi_+ \| \to 0 \quad \text{as } t \to +\infty.$$

(iv) The mapping $W_+ : \phi_+ \to \phi_0$ maps H_s into H_s and is uniformly continuous on balls in H_s.

The really difficult aspect concerns the asymptotic completeness of the wave operators $W_\pm$. It is only when asymptotic completeness is established, that is when we have proved that $R(W_+) = R(W_-)$, that we can properly define the scattering operator $S := (W_+)^{-1}W_-$.

To give some idea of what is involved let us suppose that all the hypotheses of Theorem 9.6 are satisfied so that the wave operators $W_\pm$ exist and are one to one continuous maps of H_s into itself. Now let $\phi_0 \in R(W_-)$. Theorem 9.6 then indicates that there must be a $\phi_- \in H_s$ and a solution $\phi(t) \in H$ of

$$\phi(t) = U_0(t)\phi_- + \int_{-\infty}^{t} U_0(t-s)N(\phi)(s)ds \tag{9.23}$$

such that $\phi(0) = \phi_0$ and $\| \phi(t) - U_0(t)\phi_- \| \to 0 \quad \text{as } t \to \infty$.

What we now have to do is prove that there exists an element $\phi_+ \in H_s$ such that

$$\| \phi(t) - U_0(t)\phi_+ \| \to 0 \quad \text{as } t \to +\infty. \tag{9.24}$$

If (9.24) holds then we must have

$$\phi_+ = \lim_{t \to +\infty} U_0(-t)\phi(t) \tag{9.25}$$

since $U_0(t) = \exp(-itG)$ is unitary. Consequently, a natural way to establish the existence of the limit in (9.25) is to take advantage of the closedness of H and show that $\{U_0(t)\phi(t)\}_{t\in\mathbf{R}}$ defines a Cauchy sequence in H. By (9.24) together with hypothesis **A3** we obtain

$$\begin{aligned}\| U_0(-t_1)\phi(t_1) - U_0(-t_2)\phi(t_2) \| &\leq \int_{t_2}^{t_1} \| N(\phi)(s) \| \, ds \\ &\leq \beta \int_{t_1}^{t_2} \| \phi(s) \|_1^q \| \phi(s) \| \, ds. \end{aligned} \tag{9.26}$$

We now need to show that the right hand side can become arbitrarily small as $t_1, t_2 \to +\infty$. An upper bound for $\| \phi(s) \|$ can usually be obtained, fairly readily for particular problems, from energy arguments. Thus what now remains is to show that

$$\int_{-\infty}^{\infty} \| \phi(s) \|_1^q \, ds < \infty.$$

For the particular case of the wave equation with a cubic nonlinearity

$$u_{tt} - \Delta u + m^2 u = -u^3, \quad (x,t) \in \mathbf{R}^3 \times \mathbf{R} \tag{9.27}$$

it can be proved that

$$\int_{-\infty}^{\infty} \| u(x,t) \|_\infty^2 \, dt < \infty.$$

Therefore we conclude that what is needed to settle the asymptotic completeness of the wave operators is an a priori estimate on the solution to the nonlinear equation of interest which will guarantee that solutions, for reasonable initial data, decay sufficiently rapidly in the $\| \cdot \|_1$ norm as $t \to \pm\infty$. For general non-linear problems such a priori estimates are not known although the number of results being obtained for particular problems is growing. A typical example is the following.

Theorem 9.7

Let $\mathcal{F}$ denote the closure $C_0^\infty(\mathbf{R}^3) \times C_0^\infty(\mathbf{R}^3)$ with respect to the norm

$$\begin{aligned}\| (f,g) \|_s^2 = &\sup_t \{ \int_{\mathbf{R}^3} (u_t^2 + | \Delta u |^2 + m^2 u^2) dx \} + \\ &+ \sup_t \sup_x | u(x,t) |^2 + \int_{-\infty}^{\infty} \sup_x | u(x,t) |^2 \, dt \end{aligned}$$

where $u(x,t)$ denotes the solution of the FP associated with $u_{tt} - \Delta u + m^2 u = 0$ with initial data $u(x,0) = f(x)$, $u_t(x,0) = g(x)$. The scattering operator for (9.27) exists and is a continuous mapping of $\mathcal{F}$ into itself.

Furthermore, if $(f,g) \in \mathcal{F}$ and ∇f has finite energy then the solution $u(x,t)$ of (9.27) with initial data $((f,g)$ satisfies

$$\| u(x,t) \|_\infty \leq c(1+ | t |)^{-\frac{3}{2}}.$$

Theorems 9.4, 9.5 and 9.6 provide conditions for scattering to occur. The stated results are given in forms of the norms $\| \cdot \|$, $\| \cdot \|_1$ and $\| \cdot \|_2$ and their proofs rely very much on decay estimates for the solution of the IVP in terms of these norms. Consequently, in applications a proper choice of these norms is essential. A typical strategy is to choose $\| \cdot \|_1$ as the sup. norm and $\| \cdot \|$ as the simplest Hilbert space norm that will ensure **A1** holds. The norm $\| \cdot \|_2$ is then chosen as the simplest norm for which it is possible to prove that the estimate **A2** holds. Thus we see that the three norms can be determined by considering only the linear problem, that is the FP. It now remains to show that with this choice of norms the non-linear term satisfies **A3**. If this last stage cannot be achieved then a different choice of the three norms will have to be made, provided of course that this is possible. This procedure is illustrated in the following

Example 9.8

Consider the IVP

$$\left(\frac{\partial^2}{\partial t^2} - \frac{\partial^2}{\partial x^2} + m^2\right) u(x,t) = \lambda u^p(x,t) \quad (x,t) \in \mathbf{R} \times \mathbf{R} \tag{9.28}$$

$$u(x,0) = f(x) \quad u_t(x,0) = g(x). \tag{9.29}$$

Our first requirement is a decay estimate for solutions to the linear IVP

$$\left(\frac{\partial^2}{\partial t^2} - \frac{\partial^2}{\partial x^2} + m^2\right) u(x,t) = 0 \tag{9.30}$$
$$u(x,0) = f(x) \quad u_t(x,0) = g(x). \tag{9.31}$$

The required estimate can be obtained in terms of the Sobolev- Hilbert space norm

$\| \cdot \|_{k,2}$ defined

$$\| f \|_{k,2}^2 = \sum_{|\alpha| \leq k} \| D^\alpha f \|^2, \quad k = 1, 2, ... \tag{9.32}$$

where $\| \cdot \|$ denote the usual L_2 norm. Specifically, it can be shown that if $u(x,t)$ is a solution of the linear IVP (9.30), (9.31) then

$$\| u(x,t) \|_\infty \leq ct^{-\frac{1}{2}} \{ \| f \|_{1,2} + \| f' \|_{1,2} + \| f'' \|_{1,2} + \| g \|_{1,2} + \| g' \|_{1,2} \}.$$

If we recall the notation used in arriving at (9.10), (9.11) then we see that a suitable choice of norms might be

$$\| < u, v > \|_1 := \| u \|_\infty \quad < u, v > := \begin{bmatrix} u \\ v \end{bmatrix}$$
$$\| < u, v > \|_2 := \| u \|_{1,2} + \| u' \|_{1,2} + \| u'' \|_{1,2} + \| v \|_{1,2} + \| v' \|_{1,2} .$$

For the Hilbert space H it will be convenient to take

$$H := \{ \phi = < u, v > : \| \phi \|^2 = \| A^{\frac{1}{2}} u \|_{2,2}^2 + \| v \|_{2,2}^2 < \infty \}$$

where $A = -\Delta + m^2$.

To see that with this structure **A1** holds we make use of the following consequence of the Sobolev inequalities

$$\| u \|_\infty \leq c_1 \| A^{\frac{1}{2}} u \|_{2,2}^{\frac{1}{2}} \| u \|_{2,2}^{\frac{1}{2}} \leq c_2 \| A^{\frac{1}{2}} u \|_{2,2} .$$

Since $\exp(-itA)$ is continuous and linear then

$$\| U_0(t) \phi \|_\infty \leq c_3 t^{-\frac{1}{2}} \| \phi \|^2$$

and **A2** follows.

Finally, since $N(\phi) = < 0, \lambda u^p >$ and noting the definition of H we obtain

$$\begin{aligned} \| N(\phi_1) - N(\phi_2) \| &\leq |\lambda| \, \| u_1^p - u_2^p \|_{2,2} \\ &\leq |\lambda| \, \| u_1 - u_2 \|_{2,2} \{ \| u_1 \|_\infty + \| u_2 \|_\infty \}^{p-1} \\ &\leq c |\lambda| \{ \| \phi_1 \|_1 + \| \phi_2 \|_1 \}^{p-1} \| u_1 - u_2 \| \end{aligned}$$

$$\begin{aligned}\| N(\phi_1) - N(\phi_2) \|_2 =& | \lambda | \{\| u_1^p - u_2^p \|_{1,2} + \| D(u_1^p - u_2^p) \|_{1,2}\} \\ &\leq c | \lambda | \{\| u_1 - u_2 \|_{2,2} + \| u_1' - u_2' \|_{2,2}\}\{\| u_1 \|_{2,2} + \| u_2 \|_{2,2}\}\times \\ &\times \{\| u_1 \|_\infty + \| u_2 \|_\infty\}^{p-2} \\ &\leq \beta\{\| \phi_1 \|_1 + \| \phi_2 \|_1\}^{p-2} \| \phi_1 - \phi_2 \| .\end{aligned}$$

Thus **A3** will hold provided $q = p - 2$. However, since $d = \frac{1}{2}$ and $dq > 1$ then we must choose $q > 2$ and $p > 4$.

Consequently, for the IVP (9.28), (9.29) we can obtain, via Theorems 9.4, 9.5 and 9.6, global existence of solution for small Cauchy data $< f.g >$ and the existence, of the scattering operator provided $p > 4$. This result holds irrespective of the sign of λ and no matter whether p is either positive or negative.

9.4 More on conditions ensuring scattering

A scattering state is the same notion in both linear and nonlinear systems. It is a state of the system which as $t \to \pm\infty$ behaves asymptotically like a free system, that is a system in which all interaction effects are zero. For nonlinear problems a scattering state can be thought of as one for which the inherent nonlinearity is switched off as $t \to \pm\infty$. Consequently, paralleling the strategy employed for linear problems, a scattering state for a nonlinear problem is one which as $t \to -\infty$ tends to a free state which is the solution of an associated linear problem. Recalling the notation of earlier chapters we shall denote this free solution by $U(t)g_-$ where g_- is the associated initial data. Similarly as $t \to +\infty$ this scattering state will tend to a free state denoted $U(t)g_+$ where g_+ is rarely the same as g_-.

From its definition a scattering state for a nonlinear problem must exist for all $t \in \mathbf{R}$ and must be asymptotically linear as $t \to \pm\infty$. Consequently, the assumptions necessary to ensure the existence of scattering states must be stronger than those required to guarantee just the existence of a solution to a nonlinear problem.

Since scattering is concerned with asymptotic behaviours there is a marked distinction between the scattering theories which can be developed for different equations; and this is particularly true in the nonlinear case. It turns out that the behaviour of the nonlinear term, $N(\phi)$, near $\phi = 0$ is crucial. A growth condition as $| \phi | \to \infty$ and a positivity condition will also be required to ensure that existence and uniqueness results can be obtained. Not surprisingly, therefore, statements of general results quickly become very complicated. Consequently, in order to give an indication of the types of results which can be obtained we shall confine attention in this section to the Nonlinear Schrödinger equation (NLS) for which we take

$$N(u) = | u |^{p-1} u \tag{9.33}$$

and to the Nonlinear Klein Gordon equation (NLKG) for which we take

$$N(u) = u + | u |^{p-1} u. \tag{9.34}$$

The values of the exponent p are critical in the sense that they either inhibit or permit scattering. We already have in Theorems 9.4 to 9.7 conditions under which scattering can occur. For particular problems more detailed results can be obtained. Typical of these is the following

Theorem 9.9

Let $n \geq 3$ and consider the NLS with (9.33) and the NLKG with (9.34). If

$$1 + \frac{4}{n} < p < 1 + \frac{4}{n-2}$$

then the scattering operator, S, associated with each equation maps the energy space, H_E, into itself.

However, as we have mentioned in §9.2, in order that nonlinear terms should ultimately become small, and so enable a scattering theory to be developed, it is necessary to balance the decay rates of the solutions and the degree of the nonlinearity present. In other words, if scattering states are to exist then the exponent p in (9.33) and (9.34) cannot be too small. A fundamental result in this connection is given in the following.

Theorem 9.10

Let $n \geq 1$ and consider the NLS with (9.33) and the NLKG with (9.34). Let $u(x,t)$ denote any solution with finite energy of these equations. If $| < p \leq 1 + 2/n$ then

$$\| u(t) - U(t)h \|_{H_E}$$

does **not** tend to zero as $t \to \pm\infty$ for any $h \in C_0^\infty(\mathbf{R})^n)$.

Fortunately, conditions are also available which ensure that scattering does indeed occur. As might be expected these involve decay rates of solutions and degree of nonlinearity present. Recall that when developing a scattering theory we must follow the programme of our steps listed above. Specifically, we first solve (9.13) for arbitrary data ϕ_- given at $t = -\infty$. The final step is to show that there then exists a ϕ_+ such that $\phi(t) \to U_0(t)\phi_+$ as $t \to +\infty$. To follow this programme through care is required when selecting the spaces in which to solve (9.13); appropriate decay rates are essential. In this connection convenient spaces are $L_p(L_r) := L_p(\mathbf{R}, L_r(\mathbf{R}^n))$ with norm

$$\| u \|_{p,r}^p := \int_{-\infty}^{\infty} \left\{ \int_{\mathbf{R}^n} | u(x,t) |^r \, dx \right\}^{p/r} dt. \tag{9.35}$$

When solving (9.13) we shall use the space $L_s(L_{p+1})$ with

$$s = \frac{(p-1)}{1-d}, \qquad d = \frac{n(p-1)}{2(p+1)}. \tag{9.36}$$

In this setting the following sufficient conditions for scattering can be obtained.

Theorem 9.11

Let ϕ and ϕ_- be as in (9.13) and assume that

$$1 + \frac{4}{n} \leq p \leq 1 + \frac{4}{(n-2)}, \quad n \geq 3.$$

Then there exists a $\phi_+ \in H$ as required by **Step 4** above provided:

(i) For NLS, $\phi \in L_s(L_{p+1})$ with s given by (9.36)

(ii) For NLKG, $\phi \in L_{p+1}(L_{p+1})$ with $p \leq 1 + 4/(n-1)$

(iii) For NLKG, $\phi \in L_{p+1}(W_{p+1}^{\frac{1}{2}})$ with $p > 1 + 4/(n-1)$

where W_n^m is the usual Sobolev space.

In this and previous sections we have indicated where and when solutions of a nonlinear evolution-type equations might be expected to exist. Furthermore, we have also given results which illustrate whether or or not associated scattering processes can occur. The remaining development of a scattering theory now proceeds, in principle, along similar lines to those followed for linear problems. However, the analysis is now centred on an integral equation of the form (9.13) rather than on an immediate representation of the solution as in the linear case. It is an asymptotic analysis of this nonlinear integral equation, taken together with the now well understood spectral analysis and related eigenexpansions for the **linear** operators G and A appearing in (9.9), which will furnish the required wave profiles and far field patterns. Sadly, despite this last remark many open questions still remain when attempting to develop a general scattering theory for nonlinear problems. At present more detailed results similar to those obtainable in a general linear theory have to be struggled for by working through particular problems. The analysis of these specific problems is lengthy, often quite technical, and simply beyond the scope of an introductory text such as this. However, the Commentary gives references of existing works in this connection.

Chapter 10 Commentaries

In this final chapter we provide some additional notes and remarks on the material presented in this book. We shall give typical source references together with comments on related matters. It is impossible to give credit to all sources however those cited will usually have ample additional bibliographies of their own.

Chapter 1: As its title indicates this chapter is purely introductory. Various aspects of scattering theory are presented in essentially a formal manner. Understandably the starting point is quantum scattering theory; for more on this theory see [3, 14, 55, 56, 59, 60]. Wave motion particularly on strings is developed in [5, 76]. The approach to solutions of the wave equation by considering an equivalent first order system is discussed from the standpoint of semigroup theory in [25, 50]. An investigation of solutions to the wave equation represented in the form (1.14) is fully detailed in [85]. The method of comparing solutions outlined from (1.15) onwards parallels that used so successfully in quantum scattering theory. Details of this and of their extensions to acoustic and other scattering phenomena can be found in [3, 8, 41, 55, 56, 59, 60, 85].

Chapter 2: In this chapter we gather together a number of basic mathematical 'tools' which we will require frequently in later chapters. The material is included mainly for the newcomer to the area of scattering theory who might possibly not have had the training in modern mathematical analysis which present day mathematics students receive. The account here is simply intended to provide a guide through the relevant modern functional analysis and to highlight the main results. There are a number of fine texts available which provide a thorough development of the topics introduced in this chapter and in this connection we would mention [4, 26, 27, 34, 35, 39, 46, 87]. This being said we would draw particular attention to the following important topics. The notion of completeness is crucial for many of the arguments used in developing scattering theories. A good account of the concept is given in [39] whilst fine illustrations of its use in practical situations can be found in [53, 76]. Distribution theory is comprehensively developed in [23, 35, 45, 46, 69]. We would encourage the newcomer to become familiar with distribution theory developed in the context of $\mathbf{R}^n$, $n > 1$; especially the notion of the n-dimensional Dirac delta used in Example 2.46 [6, 12, 62]. The theory of linear operators in Hilbert space is comprehensively treated in [2, 27].

Chapter 3: Here are outlined a number of the techniques which were initially developed for dealing with problems arising in quantum scattering theory and which have been successfully extended to the study of scattering problems in other fields. A comprehensive account of the quantum mechanical aspects can be found in [3, 55, 56, 59, 60].

The reduction of an initial boundary value problem for the wave equation to an initial value problem for an ordinary differential equation involving an H-valued function is an approach frequently used in semigroup theory and one which is fully treated in [25, 50, 74, 77].

The representation of a solution to **P1** given by (I) has been obtained and discussed in [85] for a number of physically significent problems in acoustics. We return to this in Chapter 7.

The reference made, when discussing **P5** and **P6**, to an energy norm is essentially an attempt to refine the solution concept mentioned in the Introduction to this chapter. We shall see later that we will be particularly interested in those solutions which have finite energy. Consequently a suitable "energy norm" must be introduced to discuss such solutions.

The illustrative example is essentially a study of a single wave process. A fully detailed, classical account of such phenomena can be found in [44]. We would emphasise that the limits indicated in (3.16) and (3.18) touch on the behaviour of the required solutions at infinity and as such they will be found to play a central role when uniqueness of solution is being discussed. An investigation of this property of a solution is frequently conducted under either one or other of the headings **radiation conditions, limiting absorption principle** and **limiting amplitude principle**; see for example [19, 20, 79].

The perturbed problem is typical of a potential scattering problem in quantum mechanics as indeed is the given account. In this connection see [3, 55, 56, 60]. The specific example has been chosen simply because we can compute all the various components required in developing a scattering theory.

Chapter 4: Spectral theory is a highly developed topic which is far reaching in its applications. Here we outline the principal notions involved when developing a scattering theory either in this or more advanced texts. Comprehensive accounts of spectral theory from a number of different standpoints are available. In this connection we would particularly mention [2, 4, 7, 16, 18, 22, 25, 27, 35, 39, 47, 56, 60, 61, 68, 87] amongst which will be found something to suit most tastes. Although the material in this chapter is quite standard we would make the following remarks. In the development given here the focus has, for ease of presentation, been largely on the features of bounded linear operators. Similar results are certainly available for unbounded operators but in this case more care is required in accounting for the domain of an operator. This aspect is exemplified in [27] and [61], the latter giving a particularly nice account of reducing subspaces.

In Theorem 4.23 there might be some anxiety about whether or not $\phi \in D(T^*)$. This can be settled by noticing that we can always write here $(\psi, 0) = 0 = ((T - \lambda I)\psi, \phi)$ which implies that $\phi \in D((T - \lambda I))$. It then follows that $(T - \lambda I)^*\phi = (T^* - \bar{\lambda})\phi = 0$ and hence $\bar{\lambda} \in \sigma_p(T^*)$.

A detailed development of measure theory with particular emphasis on applications to scattering theory can be found in [56]. The account of spectral decompositions of Hilbert spaces in [56] can be augmented by the presentations in [35, 87].

Chapter 5: One of the earliest accounts of semigroup theory is to be found in [28]. A fine introduction to the theory is given in [50]. A more advanced, modern text is [25].

Chapter 6: In this chapter we indicated some of the basic results concerning the existence, uniqueness and completeness of wave operators. In a quantum mechanical setting an immense amount of work has been done on wave operators and their properties, in this connection see [3, 7, 8, 15, 55, 60], and a comprehensive, unified presentation of the material is to be found in [56].

A detailed account of the way in which notions used in quantum mechanical potential scattering can be extended to address target scattering problems in acoustics is given in [85]; comparison results are given in [43] and the cited references

Chapter 7: An understanding of how the incident field is accounted for when developing a scattering theory is crucial. Here we show that a pulse can be conveniently approximated as a plane wave. A similar treatment of a continuous wave source is also possible [85]. An account of the retarded potential used in this approximation can be found in [6, 13, 53]. The modelling of acoustic phenomena leading to the IBVP (7.7) to (7.9) is comprehensively developed in [31]. A detailed account of radiation conditions and the concepts of incoming and outgoing waves can be found in [12, 19, 20, 79]. A good introduction to the analysis of waves on strings is given in [5], whilst a treatment of eigenexpansions and of completeness in this context can be found in [5, 13, 53, 75]. The various types of solutions which can be obtained for partial differential equations are introduced and discussed in texts dealing with the modern theories of partial differential equations; see for example [18, 21, 23, 30, 42, 86]. The notion of solutions with finite energy has been used extensively by Wilcox [85]. The reduction of an IBVP for the acoustic wave equation to an IVP for an ordinary differential equation uses techniques employed extensively in semigroup theory [25, 50, 77]. A similar reduction for IBVPs in, for instance, elastic and electromagnetic scattering is also possible [42, 43]. It should also be mentioned that a discussion of **P1** from the standpoint of Floquet theory [17, 40] provides interesting information regarding the periodicity and stability of solutions to IBVP. A detailed investigation of waves on strings using this theory is given in [84]. However, we do not develop this approach here.

Chapter 8: As this is an introductory text we have simply displayed here the more important results and concepts in order that a newcomer to the area can become familiar, as quickly as possible, with the general strategies involved in developing scattering theories. The proofs of many of the stated results are usually quite lengthy and, as they already exist in the literature cited below, can perhaps be read more profitably once the overall strategy of the subject has been appreciated. Indeed the proofs could then be usefully modified to accomodate problems other than the acoustic problem considered here [42, 43].

The material presented here is based very much on the work of Wilcox [85]. In [85] more general forms of the results stated here are proved in detail. These feature particularly problems in $\mathbf{R}^n$ and boundaries $\partial\Omega$ which are very much less ideal than those considered here. Furthermore we have really only considered here solutions in $H(\Omega)$. More general forms of solution can also be obtained; for instance, waves with finite energy are fully discussed in [85].

Other approaches to Scattering Theory can be found in [58, 78] and particularly in [41]. A fine account of abstract Scattering Theory is given in [7].

A comprehensive treatment of boundary value problems for the Helmholtz equation by integral equation methods is given in [38]. This can be used to complement the discussion given here of the waves $w_\pm$ defined in (8.47) and (8.48). A precise form of the far field amplitudes $T_\pm$ appearing in Theorem 8.6 depends, for the acoustic problems we are discussing, on the asymptotic properties of the Hankel functions which appear in the fundamental solutions of the Helmholtz equation [48]. The completeness property of the distorted waves $w_\pm$ which is implicit in the results (8.56) to (8.59) is comprehensively treated in [85].

The spectral results quoted in Theorems 12 to 14 are proved in [35].

The Limiting Absorption Principle was introduced in [19]. It has been applied to problems involving a variety of different differential expressions, boundary conditions and domains. Recently, in conjunction with the related Limiting Amplitude Principle [20] it has been successfully used to analyse scattering problems in domains involving unbounded interfaces [64, 65, 66, 67].

Chapter 9: This chapter really only gives the flavour of what an investigation of non-linear scattering problems entails. It will have become evident that associated with the development of a scattering theory for non-linear wave equations there are many unsolved problems. Those problems which have been resolved at an abstract level have almost invariably required long and highly technical proofs. It is understandable therefore that rather more headway has been made when dealing with some specific non-linear problems which arise in the applied sciences; their physical background and particular features can frequently be of considerable help in obtaining results. However, whilst the application, for example, of Theorems 9.4, 9.5 and 9.6 to specific problems is reasonably straightforward nevertheless it is always a non-trivial exercise to select the appropriate norms and obtain adequate decay estimates.

The abstract formulation of initial value problems associated with wave equations would seem to have been initiated by Segal [70] and Theorems 9.1 and 9.2 are typical of the results which were obtained. However, it should be noted that perhaps the paper which generated much of the modern day work on existence was [32]. A fine survey of local existence results can be found in [36]; see also [45] and [80]. Global existence results have been obtained by many authors and in this connection we would particularly mention [10, 24, 32, 37, 80] and the works cited in them.

Small amplitude solutions occur when, for example, perturbations are made about some known solution. Consequently, they are of interest in many areas of the applied sciences. The development of scattering theories for solutions of small amplitude was initiated by the work of Segal [71]. This work was simplified and given a more abstract setting by Strauss in [75]; here we have closely followed [75]. The plausible conditions given at the end of §9.2 are rigorously established in [37, 72]. Further useful references which contain a number of definitive results are [24, 57].

We have noted the importance of obtaining adequate, a priori estimates of the required solutions. In particular we mentioned the so-called L_∞ decay. A more detailed discussion could have been given had we been working in a general Sobolev-Hilbert space [1, 34] setting rather than the L_2 structure we have used. We would also mention that in addition to L_∞ estimates certain other estimates, known as $L_p - L_q$ estimates are also required in more detailed analyses of non-linear problems. These aspects are developed and summarised in [33, 49, 51, 52, 81, 82]. The application of these various estimates to the one-dimensional nonlinear wave equation in Example 9.8 summarises the work appearing in [75].

Finally, we would remark that in our discussion of nonlinear problems we have been concerned only with IVP. The nonlinear term has assumed the role of a potential and accordingly we have indicated the development of a potential scattering theory. Boundary conditions can in principle be absorbed into the definition of the spatial operator and a target scattering theory developed in consequence. Some results in this direction can be found in [24, 83].

References

1. Adams, R.A.: Sobolev spaces. Academic Press, New York, 1975.
2. Akhiezer, N.I. and Glaxman, I.M.: Theory of linear operators in Hilbert space Vol. I, II. Pitman-Longman, London, 1981.
3. Amrein W., Jauch, J. and Sinha, K.: Scattering theory in quantum mechanics. Benjamin-Cummings, Menlo Park, 1977.
4. Bachman, G. and Narici, L.: Functional analysis. Academic Press, New York, 1966.
5. Baldock, G.R. and Bridgeman, T.: The mathematical theory of wave motion. Ellis Horwood, Chichester, 1981.
6. Barton, G.: Elements of Green's functions and propagation. Clarendon Press, Oxford, 1989.
7. Baumgärtel, H. and Wollenberg, M.: Mathematical Scattering Theory. Operator Theory: Advances and Applications, Vol. 19, Birkhäuser-Verlag, Stuttgart, 1983.
8. Berthier, A.M.: Spectral theory and wave operators for the Schrödinger equation. Pitman Research Notes in Mathematics No. 71, Pitman, London, 1982.
9. Brenner, P. and von Wahl, W.: Global classical solutions of nonlinear wave equations. Math. Z. **176**, 87-121, 1981.
10. Browder, F.: On non-linear wave equations. Math. Z. **80**, 249-264, 1962.
11. Coddington, E. and Levinson, N.: Theory of ordinary differential equations. McGraw-Hill, New York, 1955.
12. Colton, D. and Kress, R.: Integral equation methods in scattering theory. John Wiley & Son, Chichester, 1983.

13. Courant, R. and Hilbert, D.: Methods of mathematical physics, Vol. 1, II. Wiley-Interscience, New York, 1953.

14. Cycon, H.L., Froese, R.G., Kirsh, W. and Simon, B.: Schrödinger operators with applications to quantum mechanics and global geometry. Texts and Monographs in Physics. Springer-Verlag, Berlin, 1987.

15. Dodd, R., Eilbeck, J., Gibbon, J. and Morris, H.: Solitons and nonlinear wave equations. Academic Press, New York, 1984.

16. Dunford, N. and Schwartz, J. T.: Linear Operators, Vol. I, II, III. Wiley-Interscience, New York, 1958.

17. Eastham, M.S.P.: The spectral theory of periodic differential equations. Scottish Academic Press, Edinburgh, 1973.

18. Edmunds, D.E. and Evans, W.D.: Spectral theory and differential operators. Clarendon Press, Oxford, 1989.

19. Eidus, D.M.: The principle of limiting absorption. AMS Transl., **47** (2), 157-191, 1965.

20. Eidus, D.M.: The principle of limiting amplitude. Russian Math. Surveys **24** (3), 97-167, 1969.

21. Folland, G.: Introduction to partial differential equations. Princeton University Press, Princeton, 1976.

22. Friedrichs, K.O.: Spectral theory of operators in Hilbert space. Springer-Verlag, Berlin, 1973.

23. Gelfand, I.M. and Shilov, G.E.: Generalised functions. Academic Press, New York, 1968.

24. Ginibre, J. and Velo, G.: The global Cauchy problem for the non-linear Klein-Gordon equation. Math. Z. **189**, 487-505, 1985.

25. Goldstein, J.A.: Semigroups of linear operators and applications, Oxford University Press, Oxford, 1985.

26. Griffel, D.H.: Applied functionsl analysis. Ellis Horwood, Chichester, 1981.

27. Helmberg, G.: Introduction to Spectral Theory in Hilbert Space. Elsevier, New York, 1969.

28. Hille, E. and Phillips, R.S. Functional analysis and semigroups. Amer. Math. Soc. Coll. Publ. **31**, Providence R.I., 1957.

29. Ikebe, T.: Eigenfunction expansions associated with the Schrödinger operator and their applications to scattering theory. Arch. Rational Mech. Anal. **5**, 1-34, 1960.

30. John, F.: Partial differential equations. Springer-Verlag, Berlin, 1982.

31. Jones, D.S.: Acoustic and Electromagnetic Waves. Clarendon Press, Oxford, 1986.

32. Jorgens, K.: Das Angfrangswertproblem im Grossen für eine Klasse Nichtlinear Wellengleichungen. Math. Z. **77**, 195-307, 1961.

33. Jost, R. The general theory of quantized fields. Amer. Math. Soc., Providence, R.I., 1965.

34. Kantorovich, L.V. and Akilov, G.P.: Functional analysis in normed spaces. Pergamon, Oxford, 1964.

35. Kato, T.: Perturbation theory for linear operators. Springer-Verlag, Berlin, 1966.

36. Kato, T.: Nonlinear equations of evolution in Banach spaces. Proc. Sympos. Pure Math. **45** (2), 9-23, 1986.

37. Klainerman, S.: Global existence of small amplitude solutions to nonlinear Klein-Gordon equations in four space-time dimensions. Comm. Pure Appl. Math. **38**, 631-641, 1985.

38. Kleinman, R.E. and Roach, G.F.: Boundary integral equations for the three dimensional Helmholtz equation. SIAM Reviews **16** (2), 214-236, 1974.

39. Kreyszig, E.: Introductory functional analysis with applications. John Wiley & Sons, Chichester, 1978.

40. Kuchment, P.A.: Floquet theory for partial differential equations. Russian Math. Surveys **37.4**, 1-60, 1982.

41. Lax, P.D. and Phillips, R.S.: Scattering theory. Academic Press, New York, 1967.

42. Leis, R.: Initial boundary value problems in mathematical physics. John Wiley & Son, Chichester, 1986.

43. Leis, R. and Roach, G.F.: A transmission problem for the plate equation. Proc. Roy. Soc. Edinburgh **99A**, 285-312, 1985.

44. Levin, H.: Unidirectional Wave Motions. North Holland, Amsterdam, 1978.

45. Lighthill, J.: Introduction to Fourier Analysis and generalised functions. Cambridge University Press, Cambridge, 1958.

46. Limaye, B.V.: Functional analysis. Wiley Eastern Ltd., New Delhi, 1981.

47. Lorch, E.R.: Spectral theory. Oxford University Press, Oxford, 1962.

48. Magnus, W., Oberhettinger, F. and Soni, R.P.: Formulas and theorems for special functions of mathematical physics. Springer-Verlag, Berlin, 1966.

49. Marshall, B., Strauss, W.E. and Wainger, S.: $L^p - L^q$ estimates for the Klein-Gordon equation. J. Math. Pure Appl. **59** (9), 417-440, 1980.

50. McBride, A.C.: Semigroups of linear operators: an introduction. Pitman Research Notes in Mathematics, No. 156. Pitman, London, 1987.

51. Morawetz, C.: Time decay for the nonlinear Klein-Gordon equation. Proc. Roy. Soc. Lond. **A 306**, 291-296, 1968.

52. Morawetz, C. and Strauss, W.A.: On a nonlinear scattering operator. Comm. Pure Appl. Math. **26**, 47-54, 1973.

53. Morse, P.M. and Feshbach, H.: Methods of Theoretical Physics, Vol. I, II. McGraw-Hill, New York, 1953.

54. Neittaanmaki, P. and Roach, G.F.: Weighted Sobolev spaces and exterior problems for the Helmholtz equation. Proc. Roy. Soc. Lond. **A 410**, 373-383, 1987.

55. Newton, R.G.: Scattering theory of waves and particles. McGraw-Hill, New York, 1966.

56. Pearson, D.B.: Quantum scattering and spectral theory. Academic Press, New York, 1988.

57 Pecher, H.: Nonlinear small data scattering for the wave and Klein-Gordon equation. Math. Z. **185**, 261-270, 1984.

58. Petkov, V.: Scattering theory for hyperbolic operators. Studies in mathematics and its applications, Vol. 21. North Holland, Amsterdam, 1989.

59. Prugovecki, E.: Quantum mechanics in Hilbert space. Academic Press, New York, 1981.

60. Reed, M. and Simon, B.: Methods of modern mathematical physics, Vols. 1 to IV. Academic Press, New York, 1972.

61. Riesz, F. and Sz-Nagy, B.: Functional analysis. Ungar Publ. Co., New York, 1955.

62. Roach, G.F.: Green's Functions (2nd Edition). Cambridge University Press, Cambridge, 1982.

63. Roach, G.F.: Scattering from unbounded surfaces. Proceedings Dundee Conference. Pitmans Research Notes, 248-272. Longmans, London, 1972.

64. Roach, G. F. and Zhang, B.: On Sommerfeld radiation conditions for wave propagation with two unbounded media. Proc. Roy. Soc. Edin. 149-161, 1992.

65. Roach, G. F. and Zhang, B.: A transmission problem for the reduced wave equation in inhomogeneous media with an infinite interface. Proc. Roy. Soc. Lond. **A436**, 121-140, 1992.

66. Roach, G. F. and Zhang, B.: The limiting amplitude principle for wave propagation with two unbounded media. Proc. Camb. Phil. Soc. **112**. 207-223, 1992.

67. Roach, G. F. and Zhang, B.: Spectral representations and scattering theory for the wave equation with two unbounded media. Proc. Camb. Phil. Soc. **113**, 423-447, 1993.

68 Schechter, M.: Spectra of partial differential operators. North Holland, Amsterdam, 1971

69 Schwartz, L.: Mathematics for the physical sciences. Hermann, Paris, 1966.

70. Segal, I.: Non-linear semigroups. Ann. of Math. **78** (2), 339-364, 1963.

71. Segal, I.: Dispersion for nonlinear relativistic equations, II. Ann. Sci. Ecole. Norm. Sup **I** (4), 459-497, 1968.

72. Shatan, J.: Normal forms and quadratic nonlinear Klein-Gordon equations. Comm. Pure Appl. Maths. 38, 685-696, 1985.

73 Showalter, R.E.: Hilbert space methods for partial differential equations. Pitman, London, 1977.

74. Stone, M.H.: Linear transformations in Hilbert space and their applications to analysis. Amer. Math. Soc. Coll. Publ. **15**, Providence, R.I., 1932.

75 Strauss, W.A.: Non-linear scattering theory. Scattering in Mathematical Physics, Ed. La Vita and Marchand. Reidel, Amsterdam, 53-78, 1974.

76. Strauss, W.A.: Partial differential equations: an introduction. John Wiley & Son, Chichester, 1992.

77. Tanabe, H.: Equations of Evolution. Pitman Publishing, London, 1979.

78. Taylor, J.R.: Scattering theory. Wiley, New York, 1972.

79. Vladimirov, V.S.: Equations of mathematical physics. Marcel Dekker, New York, 1971.

80. von Wahl, W.: Klassische Lösungen nichtlinearer Wellengleichungen im Grossen. Math. Z. **112**, 241-279, 1969.

81. von Wahl, W.: L^p decay rates for homogeneous wave equations. Math. Z. **120**, 93-106, 1971.

82. von Wahl, W.: Decay estimates for nonlinear wave equations. J. Func. Anal. **9**, 490-495, 1972.

83. von Wahl, W.: Regular solutions of initial boundary value problems for linear and non-linear wave equations, II. Math. Z. **142**, 121-130, 1975.

84. Werner, P.: Resonances in periodic media. Math. Meth. Applied Sci. **17**, 227-263, 1991.

85. Wilcox, C.H.: Scattering theory for the d'Alembert equation in exterior domains. Lecture Notes in Mathematics No. 442, Springer-Verlag, Berlin, 1975.

86. Wloka, J.: Partial Differential Equations. Cambridge University Press, Cambridge, 1987.

87. Yosida, K.: Functional analysis. Springer-Verlag, Berlin, 1971.

Index